CALCULUS
REFRESHER

for the

Fundamentals of Engineering Exam

Peter Schiavone, Ph.D.

Professional Publications, Inc. Belmont, California

Production Manager: Aline M. Sullivan
Acquisitions Editor: Gerald R. Galbo
Project Editor: Jessica R. Whitney-Holden
Copy Editor: Jessica R. Whitney-Holden
Book Designer: Charles P. Oey
Typesetter: Cathy Schrott
Illustrator: Cathy Schrott
Proofreader: Kate Hayes
Cover Designer: Charles P. Oey

Calculus Refresher for the Fundamentals of Engineering Exam

Printed in the United States of America

Professional Publications, Inc.
1250 Fifth Avenue, Belmont, CA 94002
(415) 593-9119
http://www.ppi2pass.com

Current printing of this edition: 1

Library of Congress Cataloging-in-Publication Data
Schiavone, Peter, 1961-
    Calculus refresher for the fundamentals of engineering exam /
    Peter Schiavone.
        p.   cm.
    ISBN 1-888577-01-0
    1. Calculus--Problems, exercises, etc.   I. Title.
QA301.S35   1997
515′.3′076--dc21
                                                96-47547
                                                CIP

Table of Contents

Preface

Over the last several years, I have been fortunate enough to have worked as a practicing engineer, as a mathematics instructor to both traditional and nontraditional students, and as a university professor in various engineering departments. These roles have given me a unique opportunity to view the teaching, learning, and application of mathematics from many different perspectives.

A few years ago, I was asked to participate in a new program offered (initially) to engineering technologists interested in upgrading to professional engineering status. The first required course in this process was calculus. Most of the students enrolled in the program had been out of school for some time, having been employed as practicing technologists. As expected, initial reactions to the course material varied widely. As the instructor charged with the responsibility of getting these people through this course, I began to take a special interest in the development of a "short, sharp, focused" way to teach calculus.

Basically, we best learn calculus (and mathematics in general) from seeing examples and practicing key concepts using a series of targeted exercises—the more exercises the better. In this publication, I have applied this understanding of learning to the list of calculus and differential equation topics required of Fundamentals of Engineering and Professional Engineering examinees. The presentation in this book is straightforward: relevant theory followed by worked examples and practice problems. The emphasis is on worked examples. There is also a supplementary appendix containing some of the more relevant precalculus techniques and necessary formulas for algebra and trigonometry.

Finally, the intention here is not to "re-teach" calculus and differential equations or to convince you that mathematics is fun, elegant, and/or relevant (I leave that to the many textbooks on the subject). Rather, the focus is on the specific topics mentioned above, with a view to developing fluency in the essential techniques using worked examples and practice problems.

Peter Schiavone, PhD
Edmonton, Alberta
Canada

Introduction

The material in this publication is based exclusively on the specific topics from calculus and from differential equations listed in the mathematics section of the *NCEES Reference Handbook*, namely (in order), differential calculus, integral calculus, centroids and moments of inertia, and differential equations. Since the NCEES contains only a list of formulas and techniques with no worked examples and very little in the way of explanatory text, the main objective of this book is to provide a collection of illustrative examples and practice problems (with full solutions in App. 1) for specific topics.

This book is divided into four main chapters and three appendices. Each topic occupies one chapter in the book, and the material is presented in the same sequence as in the NCEES, using the same notation, nomenclature, and terminology. Each chapter is further divided into one section for each of the techniques presented in the relevant part of the NCEES. Each section contains relevant theory, explanatory text, illustrative examples, and, finally, a set of practice problems. When appropriate, at the end of the chapter there is a collection of multiple-choice/FE-style exam problems based on the material covered in the chapter.

It is suggested that you focus on one chapter and one section at a time. Read the theory and explanatory text and study the worked examples before proceeding to the practice problems. The latter are an essential part of each section and have been chosen specifically to enhance your understanding of the material. You should attempt all of the practice problems at the end of each section/chapter before moving on to the next.

Solutions to every problem are included in Apps. 1 and 2. These solutions are written more to provide teaching assistance than to furnish a set of answers. You should try to follow (and mimic) the logical sequence of steps leading to the answer. This will develop your methods, techniques, and understanding of the material.

Finally, App. 3 contains a collection of relevant formulas and techniques from precalculus. This material has been selected to support the material in the chapters so that the text is essentially self-contained. In this appendix you will find trigonometric and factoring formulas, formulas for solving quadratic equations, and properties of logarithmic and exponential functions.

1

Differential Calculus

Calculus is divided into two main parts: differential calculus and integral calculus. Each part evolves from the study of a fundamental, or base, problem. This chapter is concerned with differential calculus, where the base problem is to find the slope of the line tangent to a curve at a specified point. Through the theory of limits, the solution of this base problem leads to the concept of a *derivative*, which subsequently proves to be extremely useful in a variety of other applications involving rates of change.

Consider first the base problem of differential calculus. Try to find the slope of the line tangent to the curve $y = f(x)$ at the general point $(x, y) = (x, f(x))$.

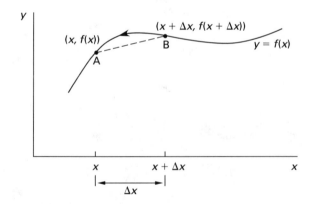

Figure 1.1 Geometric Interpretation of the Derivative

In Fig. 1.1, line AB joins points $A(x, f(x))$ and $B(x + \Delta x, f(x + \Delta x))$, where $\Delta x > 0$. The slope of line AB is given by

$$\frac{f(x + \Delta x) - f(x)}{(x + \Delta x) - x} = \frac{f(x + \Delta x) - f(x)}{\Delta x} \qquad (1.1)$$

Suppose now that B is allowed to travel along the curve toward A so that $\Delta x \to 0$. The slope of line AB approaches the slope of the curve at point $(x, f(x))$ or, more precisely, at the slope of the line tangent to curve $y = f(x)$ at point $(x, f(x))$. This can be expressed more concisely using Eq. 1.1 and the language of limits.

slope of tangent line at $(x, f(x))$

$$= \lim_{\Delta x \to 0} \frac{f(x + \Delta x) - f(x)}{\Delta x} \qquad (1.2)$$

If this limit exists, the function f is said to be *differentiable*. Since the limit in Eq. 1.2 varies with x, it defines a new function called the *derivative of f*. This is denoted by $f'(x)$ and defined by

$$f'(x) = \lim_{\Delta x \to 0} \frac{f(x + \Delta x) - f(x)}{\Delta x} \qquad (1.3)$$

Equations 1.1 through 1.3 also hold when $\Delta x < 0$ (i.e., when B lies to the left of A).

It is worth noting that Δx never actually equals zero but only tends toward zero. This ensures that points A and B remain distinct so that the slope of line AB can be computed from Eq. 1.1 no matter how close B comes to A. Consequently, the limit in Eq. 1.3 may exist even though the expression in Eq. 1.1 does not exist when $\Delta x = 0$.

Example 1.1
If $f(x) = x$, what is $f'(x)$?

Solution:

$$\frac{f(x + \Delta x) - f(x)}{\Delta x} = \frac{x + \Delta x - x}{\Delta x}$$

$$= \frac{\Delta x}{\Delta x}$$

$$= \begin{cases} 1, & \Delta x \neq 0 \\ \text{does not exist}, & \Delta x = 0 \end{cases}$$

From Eq. 1.3, the derivative of $f(x)$ is obtained as Δx *approaches* 0, not when $\Delta x = 0$. Therefore,

$$f'(x) = \lim_{\Delta x \to 0} \frac{f(x + \Delta x) - f(x)}{\Delta x}$$

$$= \lim_{\Delta x \to 0} 1$$

$$= 1$$

As expected, the slope of the line tangent to curve $y = f(x) = x$ at any point $(x, f(x))$ is 1.

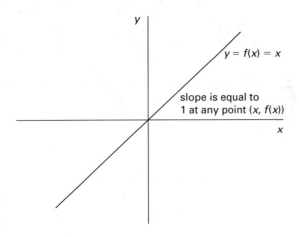

Differential calculus is concerned with the systematic study of limits such as those appearing in Eq. 1.3 (the theory of *differentiation*) and their interpretation as either slopes of tangent lines, rates of change, or functions.

1 Derivatives

There are many different notations for the first derivative of a function $y = f(x)$. These include

$$f'(x), \quad \frac{d}{dx}\big(f(x)\big), \quad \frac{df}{dx}, \quad Df(x), \quad D_x f(x), \quad D_x y, \quad y', \quad \frac{dy}{dx}$$

In most cases, the first derivative can be differentiated to obtain the second derivative. The second derivative can be denoted as follows.

$$f''(x), \quad \frac{d^2 f}{dx^2}, \quad D^2 f(x), \quad y'', \quad \frac{d^2 y}{dx^2}$$

Each notation extends naturally to third- and higher-order derivatives.

Although the definition given in Eq. 1.3 can always be used to find the derivatives of a differentiable function, it is seldom used to actually calculate derivatives because it often leads to tedious limit calculations. Over the years, systematic rules have been developed that allow the derivatives of many different types of functions to be calculated quickly and easily without recourse to the theory of limits. The derivatives of some of the more common functions arising in calculus are listed in the following table.

Table 1.1 *Derivatives of Common Functions*

$$\frac{d}{dx}(C) = 0$$

$$\frac{d}{dx}(x^n) = nx^{n-1}$$

$$\frac{d}{dx}(\sin x) = \cos x$$

$$\frac{d}{dx}(\cos x) = -\sin x$$

$$\frac{d}{dx}(\tan x) = \sec^2 x$$

$$\frac{d}{dx}(\cot x) = -\csc^2 x$$

$$\frac{d}{dx}(\sec x) = \sec x \tan x$$

$$\frac{d}{dx}(\csc x) = -\csc x \cot x$$

$$\frac{d}{dx}(\ln x) = \frac{1}{x} \, [x > 0]$$

$$\frac{d}{dx}(e^x) = e^x$$

$$\frac{d}{dx}(\arcsin x) = \frac{1}{\sqrt{1 - x^2}}$$

$$\frac{d}{dx}(\arccos x) = -\frac{1}{\sqrt{1 - x^2}}$$

$$\frac{d}{dx}(\arctan x) = \frac{1}{1 + x^2}$$

$$\frac{d}{dx}(\text{arccot}\, x) = -\frac{1}{1 + x^2}$$

$$\frac{d}{dx}(\text{arcsec}\, x) = \frac{1}{x\sqrt{1 - x^2}}$$

$$\frac{d}{dx}(\text{arccsc}\, x) = -\frac{1}{x\sqrt{1 - x^2}}$$

In Table 1.1, C and n are constants and all arguments of the trigonometric functions are in radians. Most of the functions arising in calculus are differentiated using Table 1.1 and the following rules. (In these rules, C_1 and C_2 are constants and $f(x)$ and $g(x)$ are differentiable functions of x.)

Linearity

$$\frac{d}{dx}\big((C_1 f(x) \pm C_2 g(x)\big) = C_1 \frac{d}{dx}\big(f(x)\big) \pm C_2 \frac{d}{dx}\big(g(x)\big)$$

$$(1.4)$$

Example 1.2

(a) Find $(d/dx)(3x + 2)$.

(b) Find $(d/dx)(2x^5 - 7\sin x + 3x)$.

Solution:
(a) Use Eq. 1.4 and Table 1.1.

$$\frac{d}{dx}(3x + 2) = 3\frac{d}{dx}(x) + \frac{d}{dx}(2)$$
$$= (3)(1) + (2)(0)$$
$$= 3$$

(b) Use Eq. 1.4 to split the derivative into a sum of simpler derivatives. Then use Table 1.1.

$$\frac{d}{dx}(2x^5 - 7\sin x + 3x)$$
$$= 2\frac{d}{dx}(x^5) - 7\frac{d}{dx}(\sin x) + 3\frac{d}{dx}(x)$$
$$= (2)(5x^4) - (7)(\cos x) + (3)(1)$$
$$= 10x^4 - 7\cos x + 3$$

Product Rule

$$\frac{d}{dx}[f(x)g(x)] = f(x)\frac{d}{dx}(g(x)) + g(x)\frac{d}{dx}(f(x)) \quad (1.5)$$

Example 1.3
Find $(d/dx)(e^x \cos x)$.

Solution:
Use Eq. 1.5 with $f(x) = e^x$ and $g(x) = \cos x$.

$$\frac{d}{dx}(e^x \cos x) = e^x\frac{d}{dx}(\cos x) + \cos x\frac{d}{dx}(e^x)$$
$$= e^x(-\sin x) + (\cos x)(e^x)$$
$$= e^x(\cos x - \sin x)$$

Note that the result is the same if Eq. 1.5 is used instead with $f(x) = \cos x$ and $g(x) = e^x$.

Quotient Rule

$$\frac{d}{dx}\left(\frac{f(x)}{g(x)}\right) = \frac{g(x)\frac{d}{dx}(f(x)) - f(x)\frac{d}{dx}(g(x))}{(g(x))^2} \quad (1.6)$$

Example 1.4
Calculate $(d/dx)(x^2/\sin x)$.

Solution:
Use Eq. 1.6 with $f(x) = x^2$ and $g(x) = \sin x$.

$$\frac{d}{dx}\left(\frac{x^2}{\sin x}\right) = \frac{\sin x\frac{d}{dx}(x^2) - x^2\frac{d}{dx}(\sin x)}{(\sin x)^2}$$
$$= \frac{2x\sin x - x^2\cos x}{\sin^2 x}$$

Chain Rule

Functions of functions or, more precisely, functions of the form $f(g(x))$, are differentiated as follows.

$$\frac{d}{dx}(f(g(x))) = \left[\frac{d}{dg}(f(g))\right]\left(\frac{dg}{dx}\right) \quad (1.7)$$

Alternatively, let $y = f(g(x))$ and $u = g(x)$. Then $y = f(g) = f(u)$, and Eq. 1.7 becomes

$$\frac{dy}{dx} = \left(\frac{dy}{du}\right)\left(\frac{du}{dx}\right) \quad (1.8)$$

Example 1.5
Calculate $(d/dx)(\sin(3x^2 + 2))$.

Solution:
Use Eq. 1.8 with $y = \sin(3x^2 + 2) = \sin u$, where $u = 3x^2 + 2$.

$$\frac{d}{dx}(\sin(3x^2 + 2)) = \left[\frac{d}{du}(\sin u)\right]\left(\frac{du}{dx}\right)$$
$$= \cos u\frac{d}{dx}(3x^2 + 2)$$
$$= (\cos u)(6x)$$
$$= 6x\cos(3x^2 + 2)$$

Implicit Differentiation

Implicit differentiation is a special case of the chain rule. If y is a differentiable function of x, by Eq. 1.7 with $g = y$,

$$\frac{d}{dx}(f(y(x))) = \left[\frac{d}{dy}(f(y))\right]\left(\frac{dy}{dx}\right) \quad (1.9)$$

If in Eq. 1.9 $f(y) = y^n$ where n is constant,

$$\frac{d}{dx}(y^n(x)) = \left[\frac{d}{dy}(y^n)\right]\left(\frac{dy}{dx}\right) = ny^{n-1}\frac{dy}{dx} \quad (1.10)$$

Example 1.6
(a) Determine $(d/dx)(\sin y)$ where $y(x)$ (i.e., y is a function of x).

(b) Determine $(d/dx)\sqrt{y}$ where $y(x)$ (i.e., y is a function of x).

Solution:
(a) Use Eq. 1.9 with $f(y) = \sin y$.

$$\frac{d}{dx}(\sin y) = \left(\frac{d}{dy}(\sin y)\right)\left(\frac{dy}{dx}\right) = \cos y\frac{dy}{dx}$$

(b) Use Eq. 1.10 with $n = 1/2$.

$$\frac{d}{dx}\sqrt{y} = \frac{d}{dx}\left(y^{\frac{1}{2}}\right)$$

$$= \left(\frac{d}{dy}\left(y^{\frac{1}{2}}\right)\right)\left(\frac{dy}{dx}\right)$$

$$= \frac{1}{2}y^{-\frac{1}{2}}\left(\frac{dy}{dx}\right)$$

In each case, dy/dx is left alone as part of the answer. In fact, this method is used to find dy/dx when y is given *implicitly* in terms of x.

....

Example 1.7
If $y^2 + \sin y = 4x$, what is dy/dx?

Solution:
In this case, the relation between y and x is not *explicit* but *implicit*. That is, it cannot be expressed in the form $y = f(x)$. To find dy/dx, use the following procedure, known as *implicit differentiation*.

First differentiate both sides of the equation $y^2 + \sin y = 4x$ with respect to x. Apply Eqs. 1.9 and 1.10. Finally, solve the resulting equation for dy/dx.

$$\frac{d}{dx}(y^2 + \sin y) = \frac{d}{dx}(4x)$$

$$\frac{d}{dx}(y^2) + \frac{d}{dx}(\sin y) = 4$$

$$2y\frac{dy}{dx} + \left(\frac{d}{dy}(\sin y)\right)\frac{dy}{dx} = 4$$

$$2y\frac{dy}{dx} + \cos y\frac{dy}{dx} = 4$$

Solve for dy/dx.

$$\frac{dy}{dx}(2y + \cos y) = 4$$

$$\frac{dy}{dx} = \frac{4}{2y + \cos y}$$

....

Logarithmic Differentiation
The following expression is differentiated by first simplifying using logarithms and then using Eq. 1.9. (See App. 3 for more information on logarithms.)

$$\left(f(x)\right)^{g(x)} \qquad (1.11)$$

Example 1.8
Find $(d/dx)(x^{\sin x})$.

Solution:
Let $y = x^{\sin x}$ and take logarithms of both sides.

$$y = x^{\sin x}$$

$$\ln y = \ln x^{\sin x} = \sin x \ln x$$

Using Eqs. 1.5 and 1.9, differentiate with respect to x.

$$\frac{d}{dx}(\ln y) = \frac{d}{dx}(\sin x \ln x)$$

$$\left[\frac{d}{dy}(\ln y)\right]\left(\frac{dy}{dx}\right) = (\sin x)\left(\frac{1}{x}\right) + (\cos x)(\ln x)$$

$$\left(\frac{1}{y}\right)\left(\frac{dy}{dx}\right) = \frac{\sin x}{x} + \cos x \ln x$$

$$\frac{dy}{dx} = y\left(\frac{\sin x}{x} + \cos x \ln x\right)$$

Let $y = x^{\sin x}$.

$$\frac{dy}{dx} = x^{\sin x}\left(\frac{\sin x}{x} + \cos x \ln x\right)$$

....

The procedure used in Ex. 1.8 is known as *logarithmic differentiation*. In fact, application of this procedure to the general expression in Eq. 1.11 leads to the following formula for differentiating functions of the form $\left(f(x)\right)^{g(x)}$.

$$\frac{d}{dx}\left(\left(f(x)\right)^{g(x)}\right) = g(x)[f(x)]^{g(x)-1}\frac{df}{dx}$$

$$+ [f(x)]^{g(x)}[\ln f(x)]\frac{dg}{dx} \quad (1.12)$$

One particular case of Eq. 1.12 that deserves special mention arises when $f(x) = C$, where C is a positive constant. Equation 1.12 then becomes

$$\frac{d}{dx}(C)^{g(x)} = (C)^{g(x)}(\ln C)\frac{dg}{dx} \qquad (1.13)$$

Example 1.9
Find $(d/dx)\left((\sin x)^{x^2}\right)$.

Solution:
Use Eq. 1.12 with $f(x) = \sin x$ and $g(x) = x^2$.

$$\frac{d}{dx}\left((\sin x)^{x^2}\right) = x^2(\sin x)^{x^2-1}\left[\frac{d}{dx}(\sin x)\right]$$

$$+ (\sin x)^{x^2}(\ln \sin x)(2x)$$

$$= x^2(\sin x)^{x^2-1}(\cos x)$$

$$+ (\sin x)^{x^2}(2x \ln \sin x)$$

....

Example 1.10
Calculate $(d/dx)(3^x)$.

Solution:
Use Eq. 1.13 with $C = 3$ and $g(x) = x$.

$$\frac{d}{dx}(3^x) = (3^x \ln 3)(1) = 3^x \ln 3$$

....

The following group of examples illustrate how the rules for differentiation outlined previously are used in combination with Table 1.1 to differentiate a wide variety of functions.

Example 1.11
Determine $(d/dx)\big(\cos(1-x)\big)$.

Solution:
Use Eq. 1.8 with $y = \cos(1-x) = \cos u$, where $u = 1-x$.

$$\frac{d}{dx}\big(\cos(1-x)\big) = \frac{d}{dx}(\cos u)$$
$$= \left[\frac{d}{du}(\cos u)\right]\left(\frac{du}{dx}\right)$$
$$= (-\sin u)(-1)$$
$$= \sin u$$
$$= \sin(1-x)$$

Example 1.12
Find $(d/dx)(\sin^2 x)$.

Solution:
Use Eq. 1.8 with $y = \sin^2 x = u^2$, where $u = \sin x$.

$$\frac{d}{dx}(\sin^2 x) = \frac{d}{dx}(u^2)$$
$$= \left[\frac{d}{du}(u^2)\right]\left(\frac{du}{dx}\right)$$
$$= 2u(\cos x)$$
$$= 2\sin x \cos x$$

Example 1.13
Find $(d/dx)\big((3x+2)^2 + xe^{-2x}\big)$.

Solution:
Use Eq. 1.4 to split the derivative into a sum of simpler derivatives.

$$\frac{d}{dx}\big((3x+2)^2 + xe^{-2x}\big) = \frac{d}{dx}\big((3x+2)^2\big) + \frac{d}{dx}(xe^{-2x})$$

Consider each term on the right-hand side separately. First, the term $(d/dx)\big((3x+2)^2\big)$:

Apply Eq. 1.8 with $y = (3x+2)^2 = u^2$, where $u = 3x+2$.

$$\frac{d}{dx}\big((3x+2)^2\big) = \frac{d}{dx}(u^2)$$
$$= \left[\frac{d}{du}(u^2)\right]\left(\frac{du}{dx}\right)$$
$$= 2u(3)$$
$$= 6u$$
$$= (6)(3x+2)$$

Next, consider the term $(d/dx)(xe^{-2x})$. Apply Eq. 1.5 to the term xe^{-2x}.

$$\frac{d}{dx}(xe^{-2x}) = x\frac{d}{dx}(e^{-2x}) + e^{-2x}\frac{d}{dx}(x)$$
$$= x\frac{d}{dx}(e^{-2x}) + e^{-2x}(1)$$
$$= x\frac{d}{dx}(e^{-2x}) + e^{-2x}$$

To complete this particular calculation, consider the term $(d/dx)(e^{-2x})$.

Use Eq. 1.8 with $y = e^{-2x} = e^u$, where $u = -2x$.

$$\frac{d}{dx}(e^{-2x}) = \frac{d}{dx}(e^u)$$
$$= \left[\frac{d}{du}(e^u)\right]\left(\frac{du}{dx}\right)$$
$$= e^u(-2)$$
$$= -2e^{-2x}$$

This means that

$$\frac{d}{dx}(xe^{-2x}) = -2xe^{-2x} + e^{-2x}$$

Finally, the derivative of the entire function is

$$\frac{d}{dx}\big((3x+2)^2 + xe^{-2x}\big) = (6)(3x+2) - 2xe^{-2x} + e^{-2x}$$

Example 1.14
Calculate $(d/ds)\big(\tan s/(s+1)\big)$.

Solution:
Apply Eq. 1.6 with $f(s) = \tan s$ and $g(s) = s+1$.

$$\frac{d}{ds}\left(\frac{\tan s}{s+1}\right) = \frac{(s+1)\left[\dfrac{d}{ds}(\tan s)\right] - (\tan s)\left[\dfrac{d}{ds}(s+1)\right]}{(s+1)^2}$$
$$= \frac{(s+1)\sec^2 s - (\tan s)(1)}{(s+1)^2}$$

Example 1.15
If y is given implicitly in terms of x by the relation $y^3 + \sin 2x + \ln y = 3y^2$, find dy/dx.

Solution:
Use implicit differentiation as in Ex. 1.7.

$$y^3 + \sin 2x + \ln y = 3y^2$$
$$\frac{d}{dx}\big(y^3 + \sin 2x + \ln y\big) = 3\frac{d}{dx}(y^2)$$
$$\frac{d}{dx}(y^3) + \frac{d}{dx}(\sin 2x) + \frac{d}{dx}(\ln y) = 3\frac{d}{dx}(y^2)$$

Repeated application of Eqs. 1.9 and 1.10 yields

$$\left(\frac{d}{dy}(y^3)\right)\left(\frac{dy}{dx}\right) + 2\cos 2x$$
$$+ \left(\frac{d}{dy}(\ln y)\right)\left(\frac{dy}{dx}\right) = (3)\left(\frac{d}{dy}(y^2)\right)\left(\frac{dy}{dx}\right)$$
$$3y^2\frac{dy}{dx} + 2\cos 2x + \left(\frac{1}{y}\right)\left(\frac{dy}{dx}\right) = (3)(2y)\left(\frac{dy}{dx}\right)$$

Solve for the term dy/dx.

$$\frac{dy}{dx}\left(3y^2 + \frac{1}{y} - 6y\right) = -2\cos 2x$$
$$\frac{dy}{dx} = \frac{-2\cos 2x}{3y^2 + \dfrac{1}{y} - 6y}$$

Example 1.16
Find $(d/dx)\big((x+1)^x\big)$.

Solution:
Use logarithmic differentiation as in Ex. 1.8,

$$y = (x+1)^x$$
$$\ln y = \ln(x+1)^x = x\ln(x+1)$$
$$\frac{d}{dx}(\ln y) = \frac{d}{dx}\big(x\ln(x+1)\big)$$

Apply Eq. 1.5 to the derivative on the right-hand side.

$$\frac{d}{dx}(\ln y) = x\frac{d}{dx}\big(\ln(x+1)\big) + \ln(x+1)\frac{d}{dx}(x)$$

Consider the term $(d/dx)\big(\ln(x+1)\big)$. Use Eq. 1.8 with $y = \ln(x+1) = \ln u$, where $u = x + 1$.

$$\frac{d}{dx}\big(\ln(x+1)\big) = \frac{d}{dx}(\ln u)$$
$$= \left[\frac{d}{du}(\ln u)\right]\left(\frac{du}{dx}\right)$$
$$= \left(\frac{1}{u}\right)(1)$$
$$= \frac{1}{x+1}$$

It now follows that

$$\frac{d}{dx}(\ln y) = x\left(\frac{1}{x+1}\right) + \ln(x+1)\frac{d}{dx}(x)$$
$$= \frac{x}{x+1} + [\ln(x+1)](1)$$

Apply Eq. 1.9 to the term $(d/dx)(\ln y)$.

$$\left(\frac{1}{y}\right)\left(\frac{dy}{dx}\right) = \frac{x}{x+1} + \ln(x+1)$$

Solve for dy/dx.

$$\frac{dy}{dx} = y\left[\frac{x}{x+1} + \ln(x+1)\right]$$

Let $y = (x+1)^x$.

$$\frac{dy}{dx} = (x+1)^x\left[\frac{x}{x+1} + \ln(x+1)\right]$$

Alternatively, use Eq. 1.12 with $f(x) = x+1$ and $g(x) = x$.

$$\frac{d}{dx}\big((x+1)^x\big) = x(x+1)^{x-1}$$
$$+ (x+1)^x[\ln(x+1)]\left[\frac{d}{dx}(x)\right]$$
$$= \left(\frac{x}{x+1}\right)(x+1)^x$$
$$+ (x+1)^x[\ln(x+1)](1)$$
$$= (x+1)^x\left[\frac{x}{x+1} + \ln(x+1)\right]$$

Example 1.17
Find $(d/dx)\big((\sin x)^x + e^{-8x}\big)$.

Solution:
Use Eq. 1.4 to split the derivative into a sum of two simpler derivatives.

$$\frac{d}{dx}\big((\sin x)^x + e^{-8x}\big) = \frac{d}{dx}\big((\sin x)^x\big) + \frac{d}{dx}(e^{-8x})$$

Consider each term on the right-hand side separately.

Apply Eq. 1.12 to the first term with $f(x) = \sin x$ and $g(x) = x$.

$$\frac{d}{dx}\big((\sin x)^x\big) = x(\sin x)^{x-1}\cos x + (\sin x)^x[\ln(\sin x)](1)$$

Apply Eq. 1.8 to the term $(d/dx)(e^{-8x})$ with $y = e^{-8x} = e^u$, where $u = -8x$.

$$\frac{d}{dx}(e^{-8x}) = \frac{d}{dx}(e^u)$$
$$= \left[\frac{d}{du}(e^u)\right]\left(\frac{du}{dx}\right)$$
$$= e^u(-8)$$
$$= -8e^{-8x}$$

The entire derivative is given by

$$\frac{d}{dx}\left((\sin x)^x + e^{-8x}\right) = x(\sin x)^{x-1}(\cos x)$$
$$+ (\sin x)^x \ln(\sin x) - 8e^{-8x}$$

Example 1.18

Find the slope of the line tangent to the curve with equation $y = f(x) = x^2$ at point $(1, 1)$.

Solution:

From Eqs. 1.2 and 1.3, the slope of the tangent line is given $f'(x)$.

$$y = f(x) = x^2$$
$$f'(x) = \frac{dy}{dx} = 2x$$

At point $(1, 1)$, $x = 1$, so $f'(1) = (2)(1) = 2$.

The slope of the tangent line at the point $(1, 1)$ is 2.

Example 1.19

Calculate $(d/dx)(x^7 + 3x^{\frac{1}{2}} - 2\ln(x^2 + 1))$.

Solution:

From Eq. 1.4 and Table 1.1,

$$\frac{d}{dx}\left(x^7 + 3x^{\frac{1}{2}} - 2\ln(x^2 + 1)\right)$$
$$= \frac{d}{dx}(x^7) + 3\frac{d}{dx}(x^{\frac{1}{2}}) - 2\frac{d}{dx}\left(\ln(x^2 + 1)\right)$$
$$= 7x^6 + \frac{3}{2}x^{-\frac{1}{2}} - 2\frac{d}{dx}\left(\ln(x^2 + 1)\right)$$

Consider the derivative $(d/dx)\left(\ln(x^2 + 1)\right)$. Apply Eq. 1.8 with $y = \ln(x^2 + 1) = \ln u$, where $u = x^2 + 1$.

$$\frac{d}{dx}\left(\ln(x^2 + 1)\right) = \left[\frac{d}{du}(\ln u)\right]\left(\frac{du}{dx}\right)$$
$$= \left(\frac{1}{u}\right)\left[\frac{d}{dx}(x^2 + 1)\right]$$
$$= \left(\frac{1}{u}\right)(2x)$$
$$= \frac{2x}{x^2 + 1}$$

The entire derivative is now given by

$$\frac{d}{dx}\left(x^7 + 3x^{\frac{1}{2}} - 2\ln(x^2 + 1)\right)$$
$$= 7x^6 + \frac{3}{2}x^{-\frac{1}{2}} - (2)\left(\frac{2x}{x^2 + 1}\right)$$
$$= 7x^6 + \frac{3}{2}x^{-\frac{1}{2}} - \frac{4x}{x^2 + 1}$$

Example 1.20

Find $(d/dx)\left((x\sin x + 2e^{x^2})/(3x^2 + 1)\right)$.

Solution:

Use Eq. 1.6 with $f(x) = x\sin x + 2e^{x^2}$ and $g(x) = 3x^2 + 1$.

$$\frac{d}{dx}\left(\frac{x\sin x + 2e^{x^2}}{3x^2 + 1}\right) = \frac{(3x^2 + 1)\dfrac{d}{dx}(x\sin x + 2e^{x^2})}{(3x^2 + 1)^2}$$
$$\frac{-(x\sin x + 2e^{x^2})\dfrac{d}{dx}(3x^2 + 1)}{(3x^2 + 1)^2}$$

Use Eqs. 1.4, 1.5, and 1.8 to find the derivative $d/dx(x\sin x + 2e^{x^2})$.

$$\frac{d}{dx}(x\sin x + 2e^{x^2}) = \left(x(\cos x) + (\sin x)(1)\right) + (2)(2xe^{x^2})$$

Finally,

$$\frac{d}{dx}\left(\frac{x\sin x + 2e^{x^2}}{3x^2 + 1}\right)$$
$$= \frac{(3x^2 + 1)[x(\cos x) + (\sin x)(1) + (2)(2xe^{x^2})]}{(3x^2 + 1)^2}$$
$$\frac{-(x\sin x + 2e^{x^2})(6x)}{(3x^2 + 1)^2}$$
$$= \frac{(3x^2 + 1)(x\cos x + \sin x + 4xe^{x^2})}{(3x^2 + 1)^2}$$
$$\frac{-6x(x\sin x + 2e^{x^2})}{(3x^2 + 1)^2}$$

Example 1.21

Calculate $(d/dx)\left(\arcsin(3x^2 + 4x + 2e^x) + 10^x\right)$.

Solution:

Use Eq. 1.4 to split the derivative into two simpler derivatives.

$$\frac{d}{dx}\left(\arcsin(3x^2 + 4x + 2e^x) + 10^x\right)$$
$$= \frac{d}{dx}\left(\arcsin(3x^2 + 4x + 2e^x)\right) + \frac{d}{dx}(10^x)$$

Consider first $(d/dx)\left(\arcsin(3x^2 + 4x + 2e^x)\right)$. Use Eq. 1.8 with $y = \arcsin u$, where $u = 3x^2 + 4x + 2e^x$.

$$\frac{d}{dx}\left(\arcsin(3x^2 + 4x + 2e^x)\right)$$
$$= \frac{d}{dx}(\arcsin u)$$
$$= \left[\frac{d}{du}(\arcsin u)\right]\left(\frac{du}{dx}\right)$$
$$= \left(\frac{1}{\sqrt{1 - u^2}}\right)(6x + 4 + 2e^x)$$
$$= \frac{6x + 4 + 2e^x}{\sqrt{1 - (3x^2 + 4x + 2e^x)^2}}$$

Next, consider $(d/dx)(10^x)$. Use Eq. 1.13 with $C = 10$ and $g(x) = x$.

$$\frac{d}{dx}(10^x) = 10^x \ln 10$$

The derivative of the entire function is

$$\frac{d}{dx}\left(\arcsin(3x^2 + 4x + 2e^x) + 10^x\right)$$

$$= \frac{6x + 4 + 2e^x}{\sqrt{1 - (3x^2 + 4x + 2e^x)^2}} + 10^x \ln 10$$

PRACTICE PROBLEMS

1. Find each of the following derivatives.

(a) $\dfrac{d}{dx}\left(x^{58} - 50x + \dfrac{1}{2}\right)$

(b) $\dfrac{d}{dr}\left(\dfrac{4}{3}\pi r^3\right)$

(c) $\dfrac{d}{dt}\left(\dfrac{\sqrt{10}}{t^{\frac{1}{2}}}\right)$

(d) $\dfrac{d}{dx}(x^{\frac{4}{3}} - x^{\frac{2}{3}})$

2. Find the indicated derivative.

(a) $\dfrac{d}{dt}\left(t^{\frac{1}{3}}(t + 2)\right)$

(b) $\dfrac{d}{dx}\left((x^2 + x + 1)(x^2 - 2)\right)$

(c) $\dfrac{d}{dt}\left(\dfrac{\sqrt{t} - 1}{\sqrt{t} + 1}\right)$

(d) $\dfrac{d}{dx}\left(\dfrac{1}{x^4 + x^2 - 1}\right)$

3. Find the following.

(a) $\dfrac{d}{dx}(\cos x + 2\tan x)$

(b) $\dfrac{d}{dx}\left(\dfrac{\sin x}{1 + \cos x}\right)$

(c) $\dfrac{d}{dx}(x \sin x \cos x)$

(d) $\dfrac{d}{dx}(\cos(x^3))$

(e) $\dfrac{d}{dx}\left((1 + x^{\frac{1}{2}})^{\frac{1}{3}}\right)$

(f) $\dfrac{d}{dx}(e^{x^2 + 2})$

4. Find dy/dx when y is given by the following implicit relation.

(a) $x^2 = \dfrac{y^2}{y^2 - 1}$

(b) $x^2 + y^2 = 5$

(c) $x \sin y + \sin 2y = \cos y$

5. Find $(d/dx)(x^{2x})$.

2 Critical Points

A ball thrown upward will follow a certain trajectory as it returns to earth, as shown in Fig. 1.2.

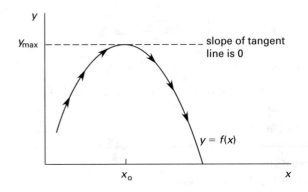

Figure 1.2 Trajectory of Ball

If the ball's trajectory is represented by the curve in Fig. 1.2, the maximum height is reached when the tangent to the curve has a slope of zero. Mathematically, if $y = f(x)$ represents the height y of the ball at position x along the ground, the ball will attain its maximum height at $x = x_o$ where, by Eqs. 1.2 and 1.3,

$$f'(x_o) = 0 \qquad (1.14)$$

Further, the maximum height is given by $y_{\max} = f(x_o)$.

Example 1.22
Let the ball's trajectory be described by the following.

$$y = f(x) = -(x - 2)^2 + 4$$

Find the ball's maximum height.

Solution:
Calculate the first derivative.

$$f'(x) = \frac{d}{dx}\left(-(x - 2)^2 + 4\right) = (-2)(x - 2)$$

Set the first derivative equal to zero as in Eq. 1.14.

$$f'(x) = (-2)(x - 2) = 0$$

This equation is satisfied when $x = 2$. The maximum height is reached when $x = x_o = 2$. The maximum height is given by

$$y_{max} = f(x_o)$$
$$= f(2)$$
$$= -(2-2)^2 + 4 = 4$$

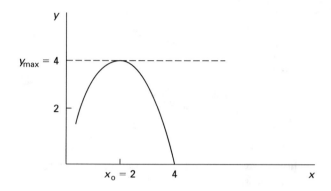

The point x_o satisfying Eq. 1.14 is called a *critical point*. It gives critical information: in this case, the maximum value of the function (height).

Derivatives can be used to locate local maximum and local minimum values of quantities represented by functions. The word *local* is used to indicate that these values may not be the absolute, or global, maximum or minimum values of the functions in an interval. In fact, global maxima or minima may occur instead at the endpoints of the interval, as illustrated in Fig. 1.3.

In general, as in Eq. 1.14, critical points are located where the first derivative is zero. That is, for a function $f(x)$, the point $x = c$ is a critical point if $f'(c) = 0$.

Example 1.23
Find the critical points of the following function.

$$f(x) = \frac{x^4}{4} - 2x^3 + \frac{11x^2}{2} - 6x$$

Solution:
Set the first derivative equal to zero. The first derivative of $f(x)$ is given by

$$f'(x) = x^3 - 6x^2 + 11x - 6$$

This equation can be factored into a product of three terms.

$$f'(x) = (x-1)(x-2)(x-3)$$

Set $f'(x)$ equal to zero.

$$(x-1)(x-2)(x-3) = 0$$

This equation is satisfied by $x = 1, 2$, or 3. The critical points are therefore $x = 1, 2, 3$.

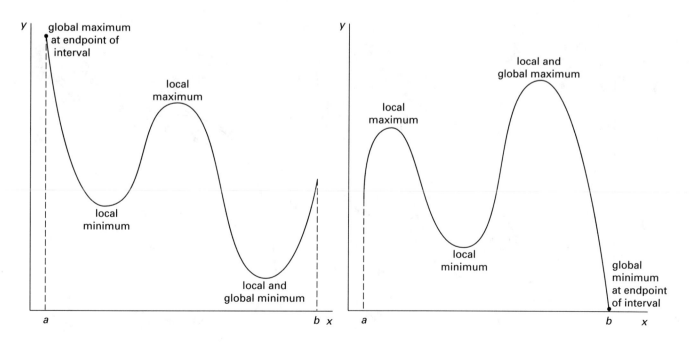

Figure 1.3 Global and Local Maxima and Minima

Once the critical points have been identified, it is necessary to decide whether each corresponds to a local maximum value or a local minimum value of the function $f(x)$. A sketch of the curve $y = f(x)$ from Ex. 1.23, indicates the following.

- $x = 1$ corresponds to a local minimum of $f(x)$.

$$y = f(1)$$
$$= \frac{(1)^4}{4} - (2)(1)^3 + \frac{(11)(1)^2}{2} - (6)(1)$$
$$= -\frac{9}{4}$$

- $x = 2$ corresponds to a local maximum of $f(x)$.

$$y = f(2)$$
$$= \frac{(2)^4}{4} - (2)(2)^3 + \frac{(11)(2)^2}{2} - (6)(2)$$
$$= -2$$

- $x = 3$ corresponds to another local minimum of $f(x)$.

$$y = f(3)$$
$$= \frac{(3)^4}{4} - (2)(3)^3 + \frac{(11)(3)^2}{2} - (6)(3)$$
$$= -\frac{9}{4}$$

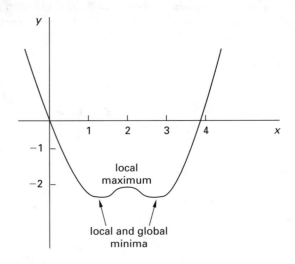

It is not always feasible (or even possible) to sketch the curve of the function $f(x)$. There is, however, a simple test for locating local maximum and local minimum values of a function. This test is based on the second derivative of the function and the notion of *concavity*. Concavity can be explained as follows.

From Eq. 1.3, the derivative $f'(x) = dy/dx$ of a function $y = f(x)$ can also be represented in the form

$$f'(x) = \frac{dy}{dx} = \lim_{\Delta x \to 0} \frac{\Delta y}{\Delta x} \qquad (1.15)$$

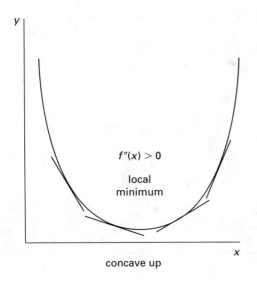

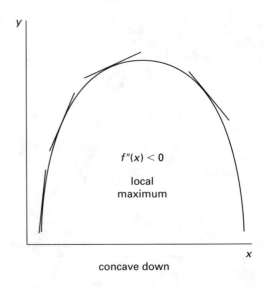

Figure 1.4 Concavity and Local Extrema

In Eq. 1.15, $\Delta y = f(x + \Delta x) - f(x)$ represents the change in the function $y = f(x)$ as the variable x changes by Δx. It follows from Eq. 1.15 that the first derivative $f'(x)$ also represents the rate of change of the function $f(x)$ with respect to x. Consequently, if $f'(x) > 0$ for all x in an interval, then the function $f(x)$ is increasing in that interval. Similarly, if $f''(x) > 0$ for all x in an interval, the function $f'(x)$ is increasing in that interval. In other words, the slope of the tangent to the curve $y = f(x)$ is increasing; the curve is *concave up* (local minimum). In the same way, if $f''(x) < 0$ for all x in an interval, the curve $y = f(x)$ is *concave down* (local maximum). Both cases are illustrated in Fig. 1.4.

The concept of concavity leads to the following simple test for locating (local) maximum and minimum values of a function.

Test 1

If, at the point $x = c$, $f'(c) = 0$ and $f''(c) > 0$, then f has a local minimum at c. If, on the other hand, $f'(c) = 0$ and $f''(c) < 0$, then f has a local maximum at c.

Note that a point on a curve is called a *point of inflection* if the curve changes concavity at that point, that is, if the second derivative changes sign at (through) that point. It follows that if there is a point of inflection at $x = c$,

$$f''(c) = 0 \qquad (1.16)$$

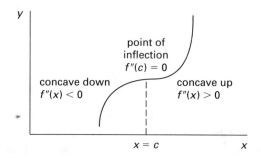

Figure 1.5 Point of Inflection

Note also that Eq. 1.16 is not sufficient to identify a point of inflection, but it is necessary, as illustrated in Ex. 1.24.

Example 1.24
Show that the function $f(x) = x^4$ does not have a point of inflection at $x = 0$ despite the fact that $f''(0) = 0$.

Solution:
If $f(x) = x^4$, then $f'(x) = 4x^3$ and $f''(x) = 12x^2$. In accordance with Eq. 1.16, set the second derivative equal to zero.

$$f''(x) = 12x^2 = 0$$

This equation is satisfied by $x = 0$. However, as shown in the following figure, there is no point of inflection at $x = 0$. The curve does not change concavity at $x = 0$ but is, in fact, concave up for all values of x.

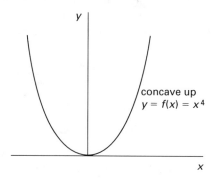

Example 1.24 illustrates that to identify $x = c$ as a point of inflection of the function $f(x)$, it is important to verify *both* of the following: $f''(c) = 0$ and $f''(x)$ changes sign at (through) $x = c$.

The following examples illustrate how to locate critical points and points of inflection and how to use Test 1 to identify local extrema.

Example 1.25
Find any local extrema (local maxima and/or local minima) and points of inflection of the function $f(x) = x^4 - 2x^2 + 3$.

Solution:
First find any critical points by setting the first derivative equal to zero.

The first derivative of $f(x)$ is given by

$$f'(x) = 4x^3 - 4x$$

This equation can be factored into a product of three factors.

$$f'(x) = 4x(x - 1)(x + 1)$$

This equation is satisfied when $x = -1$, 0, or 1. The critical points are, therefore, $x = -1$, 0, and 1. To classify each critical point as either a local maximum or local minimum, apply Test 1. The second derivative is

$$f''(x) = 12x^2 - 4$$

For the critical point $x = -1$,

$$f''(-1) = (12)(1) - 4 = 8 > 0$$

Therefore, by Test 1, the function $f(x)$ has a local minimum at $x = -1$. In fact, this local minimum is given by the value of the function $f(x)$ at $x = -1$. That is,

$$f(-1) = (-1)^4 - (2)(-1)^2 + 3 = 2$$

For the critical point $x = 0$,

$$f''(0) = (12)(0) - 4 = -4 < 0$$

Therefore, by Test 1, the function $f(x)$ has a local maximum at $x = 0$. This local maximum is given by the value of the function $f(x)$ at $x = 0$. That is,

$$f(0) = (0)^4 - (2)(0)^2 + 3 = 3$$

Finally, for the critical point $x = 1$,

$$f''(1) = (12)(1) - 4 = 8 > 0$$

Therefore, by Test 1, the function $f(x)$ has a local minimum at $x = 1$. This local minimum is given by the value of the function $f(x)$ at $x = 1$. That is,

$$f(1) = (1)^4 - (2)(1)^2 + 3 = 2$$

To identify the points of inflection, set the second derivative equal to zero.

$$12x^2 - 4 = 0$$

Factor the left-hand side of this equation.

$$(4)(3x^2 - 1) = (4)(\sqrt{3}x - 1)(\sqrt{3}x + 1) = 0$$

This equation is satisfied by $x = \pm 1/\sqrt{3}$. It follows that $f''(x) = 0$ when $x = \pm 1/\sqrt{3}$.

To determine whether these values of x correspond to points of inflection, it is necessary to check if $f''(x)$ changes sign at each value of x. Consider first $x = -1/\sqrt{3} \sim -0.577$. Pick a sample value of x slightly less than $-1/\sqrt{3}$ (e.g., $x = -1$) and one slightly greater than $-1/\sqrt{3}$ (e.g., $x = 0$). Check the sign of $f''(x)$ at each of these sample values.

$$f''(-1) = (12)(-1)^2 - 4 = 8 \quad \text{[which is positive $(+)$]}$$
$$f''(0) = (12)(0)^2 - 4 = -4 \quad \text{[which is negative $(-)$]}$$

These results are conveniently summarized in a table as follows.

sample x:	-1	$-\dfrac{1}{\sqrt{3}}$	0
sign of $f''(x)$:	$+$	0	$-$

It follows that $f''(x)$ changes sign through $x = -1/\sqrt{3}$, so $x = -1/\sqrt{3}$ does indeed correspond to a point of inflection.

Similarly, for $x = 1/\sqrt{3}$, choose sample values $x = 0$ and 1.

$$f''(0) = (12)(0)^2 - 4 = -4 \quad \text{[which is negative $(-)$]}$$
$$f''(-1) = (12)(-1)^2 - 4 = 8 \quad \text{[which is positive $(+)$]}$$

Again, there is a change in sign of $f''(x)$ through $x = 1/\sqrt{3}$. The following table summarizes the results.

sample x:	0	$\dfrac{1}{\sqrt{3}}$	1
sign of $f''(x)$:	$-$	0	$+$

The point $x = 1/\sqrt{3}$ is, therefore, also a point of inflection.

In conclusion, there are critical points at $x = -1, 0,$ and 1 (a local minimum, a local maximum, and a local minimum, respectively) and points of inflection at $x = \pm 1/\sqrt{3}$. The curve with equation $y = f(x) = x^4 - 2x^2 + 3$ is illustrated in the following figure.

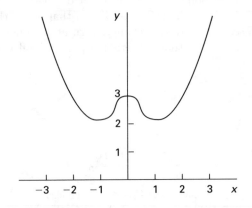

Example 1.26
Find any local extrema (local maxima and/or local minima) and points of inflection of the function $f(x) = x^3 + 3x - 10$.

Solution:
First find any critical points by setting the first derivative equal to zero. The first derivative of $f(x)$ is given by

$$f'(x) = 3x^2 + 3 = (3)(x^2 + 1)$$

This equation cannot be factored further. Set $f'(x)$ equal to zero.

$$(3)(x^2 + 1) = 0$$

Since $x^2 + 1 > 0$ for all values of x, then $f'(x) > 0$ for all values of x. Consequently, there are no values of x that make $f'(x)$ equal to zero. It follows that there are no critical points and no local extrema. The second derivative of $f(x)$ is given by

$$f''(x) = 6x$$

Set the second derivative equal to zero to identify any possible points of inflection.

$$f''(x) = 6x = 0$$

This equation is satisfied by $x = 0$. There is a possible point of inflection at $x = 0$. To check that $f''(x)$ changes sign at $x = 0$, construct a table similar to those used in Ex. 1.25, choosing appropriate sample values for x.

sample x:	-0.5	0	0.5
sign of $f''(x)$:	$-$	0	$+$

This table shows that there is indeed a point of inflection at $x = 0$. The curve $y = f(x) = x^3 + 3x - 10$, shown in the following figure, illustrates that the function is always increasing but that there is a change in concavity at the point of inflection, $x = 0$. In conclusion, $f(x)$ has a point of inflection at $x = 0$ but no local extrema.

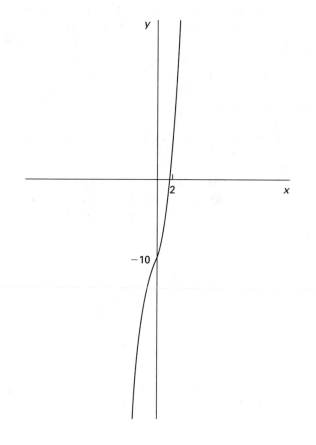

Example 1.27

Find any local extrema (local maxima and/or local minima) and points of inflection of the function $f(x) = (1/2)x - \sin x$ on the interval $0 < x < 2\pi$.

Solution:

The first and second derivatives of the function $f(x)$ are

$$f'(x) = \frac{1}{2} - \cos x$$
$$f''(x) = \sin x$$

To find critical points, set the first derivative to zero.

$$f'(x) = \frac{1}{2} - \cos x = 0$$

This equation is satisfied when $\cos x = 1/2$, which, on the interval $0 < x < 2\pi$, is satisfied by $x = \pi/3, 5\pi/3$. The critical points are, therefore, $x = \pi/3$ and $5\pi/3$. To classify the critical points as local extrema, use Test 1.

$$f''\left(\frac{\pi}{3}\right) = \sin\frac{\pi}{3} = \frac{\sqrt{3}}{2} > 0$$
$$f''\left(\frac{5\pi}{3}\right) = \sin\frac{5\pi}{3} = -\frac{\sqrt{3}}{2} < 0$$

The function $f(x)$ has a local minimum at $x = \pi/3$ and a local maximum at $x = 5\pi/3$. The local minimum has the following value.

$$f\left(\frac{\pi}{3}\right) = \left(\frac{1}{2}\right)\left(\frac{\pi}{3}\right) - \sin\frac{\pi}{3} = \frac{\pi}{6} - \frac{\sqrt{3}}{2}$$

The local maximum has the following value.

$$f\left(\frac{5\pi}{3}\right) = \left(\frac{1}{2}\right)\left(\frac{5\pi}{3}\right) - \sin\frac{5\pi}{3} = \frac{5\pi}{6} + \frac{\sqrt{3}}{2}$$

In the interval $0 < x < 2\pi$, the second derivative $f''(x) = \sin x$ is equal to zero only when $x = \pi$. Further, as in Exs. 1.25 and 1.26, the second derivative is shown to change sign at $x = \pi$, which means that there is a point of inflection at $x = \pi$ (where the curve changes concavity from the local minimum at $x = \pi/3$ to the local maximum at $x = 5\pi/3$). In conclusion, $f(x)$ has a local minimum at $x = \pi/3$, a local maximum at $x = 5\pi/3$, and a point of inflection at $x = \pi$.

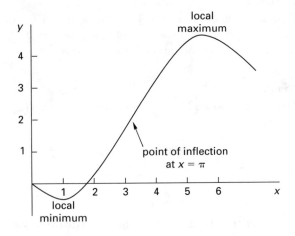

Using Eq. A, classify each critical point by evaluating the second derivative at that point and noting the sign.

$$f''(0) = (30)(0)[(2)(0) - 1] = 0$$
$$f''(1) = (30)(1)[(2)(1) - 1] = 30 > 0$$
$$f''(-1) = (30)(-1)[(2)(1) - 1] = -30 < 0$$

By Test 1, $x = 1$ corresponds to a local minimum value of $f(1) = (3)(1)^5 - (5)(1)^3 + 9 = 7$, while $x = -1$ corresponds to a local maximum value of $f(-1) = (3)(-1)^5 - (5)(-1)^3 + 9 = 11$. Test 1 gives no information on $x = 0$, but $x = 0$ might be a point of inflection. To test whether or not this is the case, form a table similar to those used in Exs. 1.25 and 1.26, choosing appropriate sample values for x.

sample x:	-0.5	0	0.5
sign of $f''(x)$:	$-$	0	$+$

It follows that $x = 0$ is neither a local maximum nor a local minimum, but rather a point of inflection. (There are two more points of inflection: $x = \pm 1/\sqrt{2}$. Each is identified, as in Exs. 1.25 and 1.26, by setting the second derivative from Eq. B equal to zero and testing for a change in sign in $f''(x)$. This identification is not important in this particular example where the emphasis is on maximum and minimum values of $f(x)$, but it would be significant when sketching the curve $y = f(x) = 3x^5 - 5x^3 + 9$.)

Finally, to find the absolute maximum and minimum values of the function $f(x)$ on the interval $[-1.5, 1.2]$, it is necessary to check the values of $f(x)$ at the endpoints of the interval.

$$f(-1.5) = (3)(-1.5)^5 - (5)(-1.5)^3 + 9 = 3.09$$
$$f(1.2) = (3)(1.2)^5 - (5)(1.2)^3 + 9 = 7.82$$

Comparing the values of the function at the endpoints with the local extrema $f(1) = 7$ and $f(-1) = 11$, the global maximum value of $f(x)$ is 11 (occurring at the local maximum $x = -1$), and the global minimum value of $f(x)$ is 3.09 (occurring at the endpoint $x = -1.5$). This is illustrated in the following figure.

Example 1.28

Find the maximum and minimum values of the function $f(x) = 3x^5 - 5x^3 + 9$ on the interval $[-1.5, 1.2]$.

Solution:

The global maximum and minimum values of the function $f(x)$ on the given interval will occur either at critical points (local extrema) or at the endpoints of the interval. First locate all critical points and find any local maxima or minima. The first derivative is

$$f'(x) = 15x^4 - 15x^2$$

This equation can be factored into a product of three factors.

$$f'(x) = 15x^2(x^2 - 1) = 15x^2(x - 1)(x + 1)$$

To find the critical points, set the first derivative equal to zero.

$$f'(x) = 15x^2(x - 1)(x + 1) = 0$$

This equation is satisfied when $x = -1$, 0, or 1. To classify the critical points, apply Test 1. The second derivative is

$$f''(x) = 60x^3 - 30x = 30x(2x^2 - 1) \qquad \text{(Eq. A)}$$

This equation can be factored into a product of three factors.

$$f''(x) = 30x(\sqrt{2}x - 1)(\sqrt{2}x + 1) \qquad \text{(Eq. B)}$$

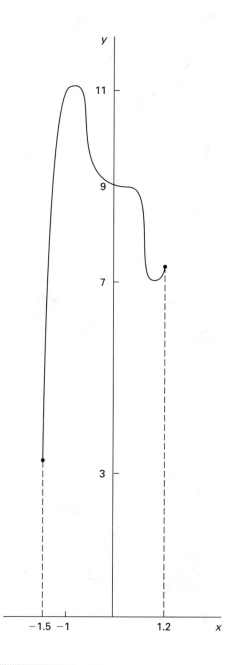

8. For the function $f(x) = x - \sqrt{2}\cos x$ on the interval $[-\pi/2, 0]$,

    (a)   locate any local maxima and/or local minima

    (b)   identify the absolute maximum and absolute minimum values

9. Find all critical points and points of inflection of the function $f(x) = x^4 - 4x^3$.

3 Partial Derivatives

Consider a function, $f(x, y)$, of two variables, x and y. If y stays fixed (e.g., $y = y_o$) and only x varies, then f behaves like a function of the single variable x. Denote this function by $g(x)$ so that $g(x) = f(x, y_o)$. In this way, Eq. 1.3 can be used to define the partial derivative of f with respect to x at (x, y_o), as follows.

$$g'(x) = f_x(x, y_o) = \lim_{\Delta x \to 0} \frac{f(x + \Delta x, y_o) - f(x, y_o)}{\Delta x}$$

As y_o changes, f_x becomes a function of the two variables x and y.

$$f_x(x, y) = \lim_{\Delta x \to 0} \frac{f(x + \Delta x, y) - f(x, y)}{\Delta x} \qquad \textit{(1.17)}$$

Equation 1.17 defines the partial derivative of the function f with respect to x.

Similarly, if Δy denotes the change in the variable y, the partial derivative of the function f with respect to y is defined by

$$f_y(x, y) = \lim_{\Delta y \to 0} \frac{f(x, y + \Delta y) - f(x, y)}{\Delta y} \qquad \textit{(1.18)}$$

To compute partial derivatives, it is seldom necessary to use Eqs. 1.17 and 1.18. Instead, it is easier to regard the partial derivative with respect to x as just the ordinary derivative of the function g of a single variable x obtained from keeping y fixed in the function $f(x, y)$. To compute f_x, y is regarded as constant, and $f(x, y)$ is differentiated with respect to x as if it were a function of the single variable x. To compute f_y, x is regarded as constant, and $f(x, y)$ is differentiated with respect to y as if it were a function of the single variable y.

In the same manner, all the rules for differentiation developed for functions of a single variable also will hold for partial differentiation of functions of more than one variable.

PRACTICE PROBLEMS

6. Find the maximum and minimum values of

    (a)   the function $f(x) = x^3 - 12x + 1$ on the interval $[-6, 4]$

    (b)   the function $f(x) = x + 1$ on the interval $[-5, 5]$

7. Find the maximum and minimum values of the function $g(x) = 2x^3 + 3x^2 + 10$ on the interval $[-2, 1]$.

As in the case of derivatives of functions of a single variable, partial derivatives can be interpreted geometrically as rates of change or as functions in their own right. For example, geometrically, in three-dimensional space, $\partial f/\partial x$ is the slope of a line tangent to the surface $z = f(x, y)$ in a plane of constant y. In terms of a rate of change, $\partial f/\partial y$ represents the rate of change of f with respect to y when x is held fixed.

There are several different notations used to denote partial derivatives. If f is a function of the two variables x and y, the partial derivative of f with respect to x is denoted by

$$f_x(x, y) = \frac{\partial f}{\partial x} = \frac{\partial}{\partial x}[f(x, y)]$$

Similarly, the partial derivative of f with respect to y is denoted by

$$f_y(x, y) = \frac{\partial f}{\partial y} = \frac{\partial}{\partial y}[f(x, y)]$$

Example 1.29

If $f(x, y) = y^3 x + x^2 y + y + x$, find $\partial f/\partial x$ and $\partial f/\partial y$ at the point $(1, 0)$.

Solution:
To find $\partial f/\partial x$, regard y as constant and differentiate f with respect to x as if it were a function of x only.

$$\frac{\partial f}{\partial x} = \frac{\partial}{\partial x}(y^3 x + x^2 y + y + x)$$

$$= \frac{\partial}{\partial x}(y^3 x) + \frac{\partial}{\partial x}(x^2 y) + \frac{\partial}{\partial x}(y) + \frac{\partial}{\partial x}(x)$$

$$= y^3 \frac{\partial}{\partial x}(x) + y\frac{\partial}{\partial x}(x^2) + \frac{\partial}{\partial x}(y) + \frac{\partial}{\partial x}(x)$$

$$= y^3 + 2xy + 0 + 1$$

$$= y^3 + 2xy + 1$$

To find $\partial f/\partial y$, regard x as constant and differentiate f with respect to y as if it were a function of y only.

$$\frac{\partial f}{\partial y} = \frac{\partial}{\partial y}(y^3 x + x^2 y + y + x)$$

$$= \frac{\partial}{\partial y}(y^3 x) + \frac{\partial}{\partial y}(x^2 y) + \frac{\partial}{\partial y}(y) + \frac{\partial}{\partial y}(x)$$

$$= x\frac{\partial}{\partial y}(y^3) + x^2 \frac{\partial}{\partial y}(y) + \frac{\partial}{\partial y}(y) + \frac{\partial}{\partial y}(x)$$

$$= 3y^2 x + x^2 + 1 + 0$$

$$= 3y^2 x + x^2 + 1$$

At the point $(1, 0)$, $x = 1$, $y = 0$, and

$$\frac{\partial f}{\partial x} = 0^3 + (2)(1)(0) + 1 = 1$$

$$\frac{\partial f}{\partial y} = (3)(0^2)(1) + 1^2 + 1 = 2$$

Example 1.30

Calculate $\partial f/\partial x$ and $\partial f/\partial y$ if $f(x, y) = y^3 \sin x + x^2 \cos y + \ln y + ye^x$.

Solution:
To find $\partial f/\partial x$, regard y as constant and differentiate f with respect to x as if it were a function of x only.

$$\frac{\partial f}{\partial x} = \frac{\partial}{\partial x}(y^3 \sin x + x^2 \cos y + \ln y + ye^x)$$

$$= \frac{\partial}{\partial x}(y^3 \sin x) + \frac{\partial}{\partial x}(x^2 \cos y) + \frac{\partial}{\partial x}(\ln y)$$

$$\quad + \frac{\partial}{\partial x}(ye^x)$$

$$= y^3 \frac{\partial}{\partial x}(\sin x) + \cos y\frac{\partial}{\partial x}(x^2) + \frac{\partial}{\partial x}(\ln y)$$

$$\quad + y\frac{\partial}{\partial x}(e^x)$$

$$= y^3 \cos x + 2x \cos y + 0 + ye^x$$

$$= y^3 \cos x + 2x \cos y + ye^x$$

To find $\partial f/\partial y$, regard x as constant and differentiate f with respect to y as if it were a function of y only.

$$\frac{\partial f}{\partial y} = \frac{\partial}{\partial y}(y^3 \sin x + x^2 \cos y + \ln y + ye^x)$$

$$= \frac{\partial}{\partial y}(y^3 \sin x) + \frac{\partial}{\partial y}(x^2 \cos y) + \frac{\partial}{\partial y}(\ln y)$$

$$\quad + \frac{\partial}{\partial y}(ye^x)$$

$$= \sin x\frac{\partial}{\partial y}(y^3) + x^2 \frac{\partial}{\partial y}(\cos y) + \frac{\partial}{\partial y}(\ln y)$$

$$\quad + e^x \frac{\partial}{\partial y}(y)$$

$$= 3y^2 \sin x - x^2 \sin y + \frac{1}{y} + e^x$$

Example 1.31

Calculate the partial derivative $\partial w/\partial y$ of the following function.

$$w(x, y) = y^2 x^3 + yx^4 + \sin y + \cos^2 y + \sin^3 x$$

Solution:
Regard x as constant.

$$\frac{\partial w}{\partial y} = \frac{\partial}{\partial y}\big(w(x,y)\big)$$

$$= \frac{\partial}{\partial y}(y^2 x^3 + yx^4 + \sin y + \cos^2 y + \sin^3 x)$$

$$= \frac{\partial}{\partial y}(y^2 x^3) + \frac{\partial}{\partial y}(yx^4) + \frac{\partial}{\partial y}(\sin y) + \frac{\partial}{\partial y}(\cos^2 y)$$
$$+ \frac{\partial}{\partial y}(\sin^3 x)$$

$$= x^3\frac{\partial}{\partial y}(y^2) + x^4\frac{\partial}{\partial y}(y) + \frac{\partial}{\partial y}(\sin y) + \frac{\partial}{\partial y}(\cos^2 y)$$
$$+ \frac{\partial}{\partial y}(\sin^3 x)$$

$$= x^3(2y) + x^4(1) + \cos y + 2\cos y(-\sin y) + 0$$

$$= 2yx^3 + x^4 + \cos y - 2\sin y\cos y$$

In the following examples, certain rules for differentiating functions of one variable (Eqs. 1.4 through 1.9) are adapted to partial differentiation. The rules themselves do not change, but any differentiation is interpreted in the sense of partial differentiation.

Example 1.32
Find $\partial z/\partial x$ and $\partial z/\partial y$ if $z = g(x,y) = \cos\big(y/(1+x)\big)$.

Solution:
To find $\partial z/\partial x$, regard y as constant so that $z = g(x,y)$ can be differentiated as if it were a function of the single variable x. With $z = \cos u$ where $u = y/(1+x)$, Eq. 1.8 can be written as

$$\frac{\partial z}{\partial x} = \left(\frac{dz}{du}\right)\left(\frac{\partial u}{\partial x}\right)$$

$$= (-\sin u)\frac{\partial}{\partial x}\left(\frac{y}{1+x}\right)$$

To find $(\partial/\partial x)\big(y/(1+x)\big)$, use Eq. 1.6.

$$\frac{\partial}{\partial x}\left(\frac{y}{1+x}\right) = \frac{(1+x)\dfrac{\partial}{\partial x}(y) - y\dfrac{\partial}{\partial x}(1+x)}{(1+x)^2}$$

$$= \frac{(1+x)(0) - y(1)}{(1+x)^2} = \frac{-y}{(1+x)^2}$$

The entire partial derivative is now given by

$$\frac{\partial z}{\partial x} = \left(\frac{dz}{du}\right)\left(\frac{\partial u}{\partial x}\right)$$

$$= (-\sin u)\left[\frac{-y}{(1+x)^2}\right]$$

$$= \left[\frac{y}{(1+x)^2}\right]\sin u$$

$$= \left[\frac{y}{(1+x)^2}\right]\sin\left(\frac{y}{1+x}\right)$$

Similarly,

$$\frac{\partial z}{\partial y} = \left(\frac{dz}{du}\right)\left(\frac{\partial u}{\partial y}\right)$$

$$= (-\sin u)\frac{\partial}{\partial y}\left(\frac{y}{1+x}\right)$$

$$= (-\sin u)\left(\frac{1}{1+x}\right)\frac{\partial}{\partial y}(y)$$

$$= (-\sin u)\left(\frac{1}{1+x}\right)(1)$$

$$= \left(-\frac{1}{1+x}\right)\sin\left(\frac{y}{1+x}\right)$$

Example 1.33
Let $z = f(x,y)$ be defined implicitly by the following equation.

$$x^2 + x\sin y + z^2 + 3xyz = 1$$

Find $\partial z/\partial x$ and $\partial z/\partial y$.

Solution:
To find $\partial z/\partial x$, differentiate implicitly (partially) with respect to x using the procedure developed in Ex. 1.7.

$$x^2 + x\sin y + z^2 + 3xyz = 1$$

Differentiate partially both sides of this equation with respect to x.

$$\frac{\partial}{\partial x}(x^2 + x\sin y + z^2 + 3xyz) = \frac{\partial}{\partial x}(1)$$

$$\frac{\partial}{\partial x}(x^2) + \frac{\partial}{\partial x}(x\sin y) + \frac{\partial}{\partial x}(z^2) + \frac{\partial}{\partial x}(3xyz) = 0$$

Regard y as constant.

$$\frac{\partial}{\partial x}(x^2) + \sin y\frac{\partial}{\partial x}(x) + \frac{\partial}{\partial x}(z^2) + 3y\frac{\partial}{\partial x}(xz) = 0 \quad \text{(Eq. A)}$$

To obtain $(\partial/\partial x)(z^2)$, apply Eq. 1.10 noting that here, z is regarded as a function of x only (y is regarded as constant).

$$\frac{\partial}{\partial x}(z^2) = 2z\frac{\partial z}{\partial x}$$

For the term $(\partial/\partial x)(xz)$ in Eq. A, again regard z as a function of x only and apply Eq. 1.5.

$$\frac{\partial}{\partial x}(xz) = x\frac{\partial}{\partial x}(z) + z\frac{\partial}{\partial x}(x)$$

$$= x\frac{\partial z}{\partial x} + z(1)$$

$$= x\frac{\partial z}{\partial x} + z$$

Returning to Eq. A,

$$2x + \sin y(1) + 2z\frac{\partial z}{\partial x} + 3y\left(x\frac{\partial z}{\partial x} + z\right) = 0$$

Solve for the term $\partial z/\partial x$.

$$\frac{\partial z}{\partial x}(2z + 3xy) = -(2x + \sin y + 3yz)$$

$$\frac{\partial z}{\partial x} = -\frac{2x + \sin y + 3yz}{2z + 3xy}$$

To find $\partial z/\partial y$, differentiate implicitly (partially) with respect to y.

$$x^2 + x\sin y + z^2 + 3xyz = 1$$

Differentiate partially both sides of this equation with respect to y.

$$\frac{\partial}{\partial y}(x^2 + x\sin y + z^2 + 3xyz) = \frac{\partial}{\partial y}(1)$$

$$\frac{\partial}{\partial y}(x^2) + \frac{\partial}{\partial y}(x\sin y) + \frac{\partial}{\partial y}(z^2) + \frac{\partial}{\partial y}(3xyz) = 0$$

Now regard x as constant.

$$\frac{\partial}{\partial y}(x^2) + x\frac{\partial}{\partial y}(\sin y) + \frac{\partial}{\partial y}(z^2) + 3x\frac{\partial}{\partial y}(yz) = 0 \quad \text{(Eq. B)}$$

To obtain $(\partial/\partial y)(z^2)$, follow the same procedure used to obtain $(\partial/\partial x)(z^2)$.

Apply Eq. 1.10 noting that now, z is regarded as a function of y only.

$$\frac{\partial}{\partial y}(z^2) = 2z\frac{\partial z}{\partial y}$$

For the term $(\partial/\partial y)(yz)$, follow the same procedure used to obtain $(\partial/\partial x)(xz)$. Regard z as a function of y only and apply Eq. 1.5.

$$\frac{\partial}{\partial y}(yz) = y\frac{\partial}{\partial y}(z) + z\frac{\partial}{\partial y}(y)$$

$$= y\frac{\partial z}{\partial y} + z(1)$$

$$= y\frac{\partial z}{\partial y} + z$$

Returning to Eq. B,

$$0 + x(\cos y) + 2z\frac{\partial z}{\partial y} + 3x\left(z + y\frac{\partial z}{\partial y}\right) = 0$$

Solve for the term $\partial z/\partial y$.

$$\frac{\partial z}{\partial y}(2z + 3xy) = -(x\cos y + 3xz)$$

$$\frac{\partial z}{\partial y} = -\frac{x\cos y + 3xz}{2z + 3xy}$$

Functions of More than Two Variables

The definition of a partial derivative extends readily to functions of three or more variables. For example, if f is a function of the three variables x, y, and z and Δz represents the change in the variable z, the partial derivative of $f(x, y, z)$ with respect to z is defined as

$$\frac{\partial f}{\partial z} = f_z(x, y, z) = \lim_{\Delta z \to 0} \frac{f(x, y, z + \Delta z) - f(x, y, z)}{\Delta z}$$

In practice, this partial derivative is found by regarding x and y as constant and differentiating $f(x, y, z)$ as if it were a function of z only. The partial derivative $\partial f/\partial x$ is found by regarding y and z as constant and by differentiating f as if it were a function of x only. The partial derivative $\partial f/\partial y$ is found by regarding x and z as constant and differentiating f as if it were a function of y only.

Example 1.34

Find $\partial f/\partial x$, $\partial f/\partial y$, and $\partial f/\partial z$ for the following function.

$$f(x, y, z) = x^2 yz + z^2 + y\sin x + xe^z$$

Solution:

For $\partial f/\partial x$, regard y and z as constant and differentiate with respect to x.

$$\frac{\partial f}{\partial x} = \frac{\partial}{\partial x}(x^2 yz + z^2 + y\sin x + xe^z)$$

$$= \frac{\partial}{\partial x}(x^2 yz) + \frac{\partial}{\partial x}(z^2) + \frac{\partial}{\partial x}(y\sin x) + \frac{\partial}{\partial x}(xe^z)$$

$$= yz\frac{\partial}{\partial x}(x^2) + \frac{\partial}{\partial x}(z^2) + y\frac{\partial}{\partial x}(\sin x) + e^z\frac{\partial}{\partial x}(x)$$

$$= yz(2x) + 0 + y\cos x + e^z(1)$$

$$= 2xyz + y\cos x + e^z$$

For $\partial f\partial y$, regard x and z as constant and differentiate with respect to y.

$$\frac{\partial f}{\partial y} = \frac{\partial}{\partial y}(x^2 yz + z^2 + y\sin x + xe^z)$$

$$= \frac{\partial}{\partial y}(x^2 yz) + \frac{\partial}{\partial y}(z^2) + \frac{\partial}{\partial y}(y\sin x) + \frac{\partial}{\partial y}(xe^z)$$

$$= x^2 z\frac{\partial}{\partial y}(y) + \frac{\partial}{\partial y}(z^2) + \sin x\frac{\partial}{\partial y}(y) + xe^z\frac{\partial}{\partial y}(1)$$

$$= x^2 z(1) + 0 + (\sin x)(1) + 0$$

$$= x^2 z + \sin x$$

For $\partial f/\partial z$, regard x and y as constant and differentiate with respect to z.

$$\frac{\partial f}{\partial z} = \frac{\partial}{\partial z}(x^2yz + z^2 + y\sin x + xe^z)$$

$$= \frac{\partial}{\partial z}(x^2yz) + \frac{\partial}{\partial z}(z^2) + \frac{\partial}{\partial z}(y\sin x) + \frac{\partial}{\partial z}(xe^z)$$

$$= x^2y\frac{\partial}{\partial z}(z) + \frac{\partial}{\partial z}(z^2) + (y\sin x)\frac{\partial}{\partial z}(1) + x\frac{\partial}{\partial z}(e^z)$$

$$= x^2y(1) + 2z + (y\sin x)0 + xe^z$$

$$= x^2y + 2z + xe^z$$

Note the essential difference in the variable z as used in Exs. 1.33 and 1.34. In Ex. 1.33, z depends on x and y so that $\partial z/\partial x$ and $\partial z/\partial y$ are not zero. In Ex. 1.34, z is an independent (of x and y) variable, implying that $\partial z/\partial x$ and $\partial z/\partial y$ are both zero.

Example 1.35
Find $\partial f/\partial x$ if $f(x, y, z) = x^2y^2z^2$.

Solution:
Regard y and z as constant and differentiate f with respect to x.

$$\frac{\partial f}{\partial x} = \frac{\partial}{\partial x}(x^2y^2z^2)$$

$$= y^2z^2\frac{\partial}{\partial x}(x^2)$$

$$= y^2z^2(2x) = 2xy^2z^2$$

Example 1.36
Find $\partial f/\partial y$ if $f(x, y, z) = \sqrt{x + y + z}$.

Solution:
Regarding x and z as constant,

$$\frac{\partial f}{\partial y} = \frac{\partial}{\partial y}\sqrt{x + y + z}$$

Apply Eq. 1.8 with $u = x + y + z$.

$$\frac{\partial f}{\partial y} = \frac{\partial}{\partial y}\sqrt{x + y + z}$$

$$= \frac{d}{du}\sqrt{u}\left(\frac{\partial u}{\partial y}\right)$$

$$= \left(\frac{1}{2}u^{-\frac{1}{2}}\right)\frac{\partial}{\partial y}(x + y + z)$$

$$= \left(\frac{1}{2}u^{-\frac{1}{2}}\right)(0 + 1 + 0)$$

$$= \frac{1}{2\sqrt{u}}$$

$$= \frac{1}{2\sqrt{x + y + z}}$$

Higher Derivatives

The above examples demonstrate that, in general, if $f(x, y)$ is a function of two variables, then so are its derivatives, f_x and f_y. These can then be differentiated to give second partial derivatives.

$$(f_x)_x = f_{xx} = \left(\frac{\partial}{\partial x}\right)\left(\frac{\partial f}{\partial x}\right) = \frac{\partial^2 f}{\partial x^2}$$

$$(f_y)_y = f_{yy} = \left(\frac{\partial}{\partial y}\right)\left(\frac{\partial f}{\partial y}\right) = \frac{\partial^2 f}{\partial y^2}$$

There are also *mixed derivatives* to consider.

$$(f_x)_y = f_{xy} = \left(\frac{\partial}{\partial y}\right)\left(\frac{\partial f}{\partial x}\right) = \frac{\partial^2 f}{\partial y\partial x}$$

$$(f_y)_x = f_{yx} = \left(\frac{\partial}{\partial x}\right)\left(\frac{\partial f}{\partial y}\right) = \frac{\partial^2 f}{\partial x\partial y}$$

Partial derivatives of order three and higher can be defined similarly. For example,

$$f_{yxx} = (f_{yx})_x = \left(\frac{\partial}{\partial x}\right)\left(\frac{\partial^2 f}{\partial y\partial x}\right) = \frac{\partial^3 f}{\partial y\partial x^2}$$

Example 1.37
Find the second partial derivatives of $g(x, y) = y^2x + \cos(x + y)$.

Solution:
Begin by finding the first partial derivatives of g.

$$\frac{\partial g}{\partial x} = \frac{\partial}{\partial x}\left(y^2x + \cos(x + y)\right)$$

$$= y^2\frac{\partial}{\partial x}(x) + \frac{\partial}{\partial x}\left(\cos(x + y)\right)$$

For $(\partial/\partial x)\left(\cos(x + y)\right)$, use Eq. 1.8 with $u = x + y$. That is,

$$\frac{\partial}{\partial x}\left(\cos(x + y)\right) = \left(\frac{d}{du}(\cos u)\right)\left(\frac{\partial u}{\partial x}\right)$$

$$= (-\sin u)(1)$$

$$= -\sin(x + y)$$

Finally,

$$\frac{\partial g}{\partial x} = y^2(1) - \sin(x + y)$$

$$= y^2 - \sin(x + y)$$

Similarly,

$$\frac{\partial g}{\partial y} = \frac{\partial}{\partial y}\left(y^2x + \cos(x + y)\right)$$

$$= x\frac{\partial}{\partial y}(y^2) + \frac{\partial}{\partial y}\left(\cos(x + y)\right)$$

$$= 2xy - \sin(x + y)$$

The second partial derivatives are obtained by differentiating the first partial derivatives.

$$\frac{\partial^2 g}{\partial x^2} = \left(\frac{\partial}{\partial x}\right)\left(\frac{\partial g}{\partial x}\right) = \frac{\partial}{\partial x}\left(y^2 - \sin\left(x+y\right)\right)$$

$$= \frac{\partial}{\partial x}(y^2) - \frac{\partial}{\partial x}\left(\sin\left(x+y\right)\right)$$

$$= 0 - \cos\left(x+y\right) = -\cos\left(x+y\right)$$

Similarly,

$$\frac{\partial^2 g}{\partial y^2} = \left(\frac{\partial}{\partial y}\right)\left(\frac{\partial g}{\partial y}\right)$$

$$= \frac{\partial}{\partial y}\left(2xy - \sin\left(x+y\right)\right)$$

$$= 2x - \cos\left(x+y\right)$$

The mixed derivatives are obtained as follows.

$$\frac{\partial^2 g}{\partial y \partial x} = \left(\frac{\partial}{\partial y}\right)\left(\frac{\partial g}{\partial x}\right) = \frac{\partial}{\partial y}\left(y^2 - \sin\left(x+y\right)\right)$$

$$= 2y - \cos\left(x+y\right)$$

$$\frac{\partial^2 g}{\partial x \partial y} = \left(\frac{\partial}{\partial x}\right)\left(\frac{\partial g}{\partial y}\right) = \frac{\partial}{\partial x}\left(2xy - \sin\left(x+y\right)\right)$$

$$= 2y - \cos\left(x+y\right)$$

Notice that $\partial^2 g/\partial x \partial y = \partial^2 g/\partial y \partial x$. This is true for most functions occurring in practice.

....

PRACTICE PROBLEMS

10. Find $\partial f/\partial x$ and $\partial f/\partial y$ when

    (a) $f(x,y) = x^3 y^2 + yx^4 + \sin y + \cos^2 y + \sin^3 x$

    (b) $f(x,y) = \ln\left(xy\right) + \dfrac{x}{y}$

    (c) $f(x,y) = \dfrac{x^3 - y^2}{x+y}$

11. Find $\partial f/\partial x$, $\partial f/\partial y$, and $\partial f/\partial z$.

    (a) $f(x,y,z) = z\ln\left(x+y+z\right)$

    (b) $f(x,y,z) = 3y^2 + 6xz + \dfrac{x}{\ln z} + \sin\left(x^2 + z\right)$

12. Find the second partial derivatives of

$$G(x,y) = \ln\left(x+y\right) - \sin\left(x-y\right) + e^{xy}$$

Verify that $G_{xy} = G_{yx}$.

4 Curvature

The *curvature* at a point A of a curve is defined as

$$K = \left|\frac{d\alpha}{ds}\right| \qquad\qquad (1.19)$$

In Eq. 1.19, α is the angle of inclination of the tangent line at A, and s is the arc length, or the distance traveled along the curve, as shown in Fig. 1.6.

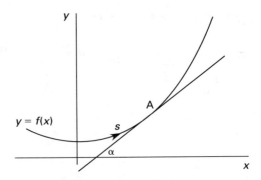

Figure 1.6 Curvature Defined in Terms of the Angle α and Arc Length s

From Eq. 1.19, it follows that the curvature is the absolute value of the rate of change of the angle α with respect to arc length s. In other words, the curvature is a measure of the rate of change of direction (sharpness) of the curve at A.

Suppose the curve in question has equation $y = f(x)$. It is possible to derive a formula for the curvature K in terms of y and its derivatives. From Eq. 1.8,

$$\frac{d\alpha}{ds} = \left(\frac{d\alpha}{dx}\right)\left(\frac{dx}{ds}\right) \qquad\qquad (1.20)$$

The slope of the line tangent at A is given by

$$y' = \tan\alpha$$

This allows us to write $d\alpha/dx$ in terms of the derivatives of y.

$$\alpha = \arctan\left(y'\right)$$

$$\frac{d\alpha}{dx} = \frac{d}{dx}\left(\arctan\left(y'\right)\right)$$

Applying Eq. 1.8 with $u = y'$,

$$\frac{d\alpha}{dx} = \left[\frac{d}{du}\left(\arctan u\right)\right]\left(\frac{du}{dx}\right)$$

$$= \left(\frac{1}{1+u^2}\right)\left(\frac{d}{dx}(y')\right)$$

$$= \left(\frac{1}{1+u^2}\right)(y'')$$

Finally, with $u = y'$,

$$\frac{d\alpha}{dx} = \left(\frac{1}{1+(y')^2}\right)(y'') = \frac{y''}{1+(y')^2} \qquad (1.21)$$

Next, with Eq. 1.20 in mind, derive an expression for dx/ds in terms of the derivatives of y. From Fig. 1.7, as Δs approaches zero,

$$\Delta s = \sqrt{(\Delta x)^2 + (\Delta y)^2} \qquad (1.22)$$

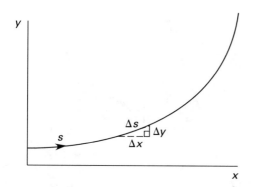

Figure 1.7 Δs *Expressed in Terms of* Δx *and* Δy

From Eq. 1.22,

$$\frac{\Delta s}{\Delta x} = \frac{\sqrt{(\Delta x)^2 + (\Delta y)^2}}{\Delta x} = \sqrt{1 + \left(\frac{\Delta y}{\Delta x}\right)^2} \qquad (1.23)$$

As Δx approaches zero, $\Delta s/\Delta x$ approaches ds/dx and $\Delta y/\Delta x$ approaches $dy/dx = y'$. From Eq. 1.23, it follows that as Δs and Δx both approach zero,

$$\frac{ds}{dx} = \sqrt{1 + \left(\frac{dy}{dx}\right)^2}$$

This equation leads to Eq. 1.24.

$$\frac{dx}{ds} = \frac{1}{\dfrac{ds}{dx}}$$

$$= \frac{1}{\sqrt{1 + \left(\dfrac{dy}{dx}\right)^2}}$$

$$= \frac{1}{\sqrt{1 + (y')^2}} \qquad (1.24)$$

From Eqs. 1.20, 1.21, and 1.24, we obtain a formula for $d\alpha/ds$.

$$\frac{d\alpha}{ds} = \left(\frac{d\alpha}{dx}\right)\left(\frac{dx}{ds}\right)$$

$$= \left[\frac{y''}{1 + (y')^2}\right]\left[\frac{1}{\sqrt{1 + (y')^2}}\right]$$

$$= \frac{y''}{[1 + (y')^2]^{\frac{3}{2}}}$$

Finally, from Eq. 1.19, for the curve $y = f(x)$, the curvature K at any point (x, y) is given by

$$K = \left|\frac{y''}{[1 + (y')^2]^{\frac{3}{2}}}\right| \qquad (1.25)$$

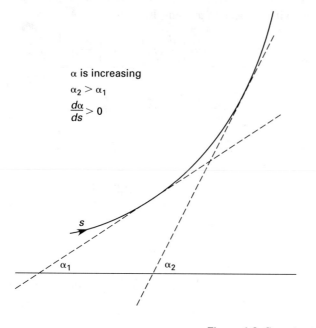
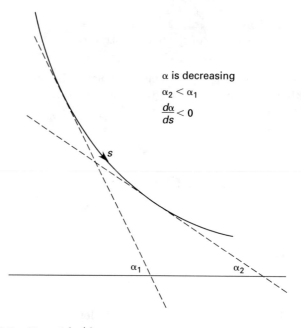

Figure 1.8 *Curvature and the Sign of* $d\alpha/ds$

Suppose the equation of the curve is written as $x = g(y)$. Using the same procedure that led to Eq. 1.25, K is expressed in terms of x and its derivatives.

$$K = \left| \frac{x''}{[1 + (x')^2]^{\frac{3}{2}}} \right| \qquad (1.26)$$

In Eq. 1.26, $x' = dx/dy$ since differentiation is now with respect to y.

Figure 1.8 illustrates that when traveling along the curve in the direction of increasing s, $d\alpha/ds$ is positive (α is increasing) where the curve bends to the left. In this case, the absolute value signs become redundant in Eqs. 1.19 and 1.25.

$$K = \frac{d\alpha}{ds} = \frac{y''}{\left(1 + (y')^2\right)^{\frac{3}{2}}}$$

Similarly, $d\alpha/ds$ is negative (α is decreasing) where the curve bends to the right. Equations 1.19 and 1.25 lead to the following expression.

$$K = -\frac{d\alpha}{ds} = -\frac{y''}{\left(1 + (y')^2\right)^{\frac{3}{2}}}$$

(Similar statements can be made regarding Eq. 1.26.)

If a curve has curvature $K \neq 0$ at a point (x, y), then the *radius of curvature* of the curve at this point is defined by

$$R = \frac{1}{K}$$

Example 1.38
Find the curvature of the line $y = 3x + 2$.

Solution:
From the following figure, since the angle α is constant with respect to the arc length s, the curvature is zero. This can be verified using Eq. 1.25,

$$y' = \frac{d}{dx}(3x + 2) = 3$$

$$y'' = \frac{d}{dx}(y') = \frac{d}{dx}(3) = 0$$

From Eq. 1.25,

$$K = \left| \frac{0}{[1 + (0)^2]^{\frac{3}{2}}} \right| = 0$$

It follows that the curvature K is zero for any point on the line.

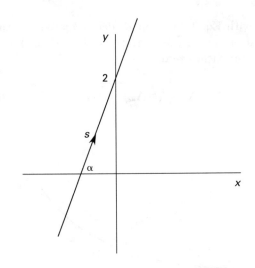

Example 1.39
Find the curvature of the circle $x^2 + y^2 = r^2$ (r^2 is a given constant).

Solution:
From the following figure, the circle changes direction at the same rate for all values of x. This suggests that the circle has constant curvature. In fact, differentiating implicitly with respect to x (as in Ex. 1.7) and using Eq. 1.10,

$$x^2 + y^2 = r^2$$

$$\frac{d}{dx}(x^2 + y^2) = \frac{d}{dx}(r^2)$$

$$2x + 2yy' = 0$$

$$y' = -\frac{x}{y}$$

The second derivative, y'', is obtained by differentiating y' with respect to x using Eq. 1.6.

$$y'' = -\frac{(y)\left[\dfrac{d}{dx}(x)\right] - (x)\left[\dfrac{d}{dx}(y)\right]}{y^2}$$

$$= -\frac{(y)(1) - (x)(y')}{y^2}$$

Substitute $y' = -x/y$ to obtain the following.

$$y'' = -\frac{y - (x)\left(-\dfrac{x}{y}\right)}{y^2}$$

$$= -\frac{y^2 + x^2}{y^3}$$

Finally, let $x^2 + y^2 = r^2$. The second derivative then becomes

$$y'' = -\frac{r^2}{y^3}$$

From Eq. 1.25,

$$K = \left| \frac{y''}{[1+(y')^2]^{\frac{3}{2}}} \right|$$

$$= \left| \frac{\left(-\dfrac{r^2}{y^3}\right)}{\left[1+\left(-\dfrac{x}{y}\right)^2\right]^{\frac{3}{2}}} \right|$$

$$= \left| \frac{\left(-\dfrac{r^2}{y^3}\right)}{\left[\dfrac{y^2+x^2}{y^2}\right]^{\frac{3}{2}}} \right|$$

Recall that $x^2 + y^2 = r^2$.

$$K = \left| \frac{\left(-\dfrac{r^2}{y^3}\right)}{\left[\dfrac{r^2}{y^2}\right]^{\frac{3}{2}}} \right|$$

$$= \left| \frac{\left(-\dfrac{r^2}{y^3}\right)}{\left[\dfrac{r^3}{y^3}\right]} \right|$$

$$= \left| -\frac{1}{r} \right|$$

$$= \frac{1}{r}$$

The curvature $K = 1/r$ is constant for every point on the circle.

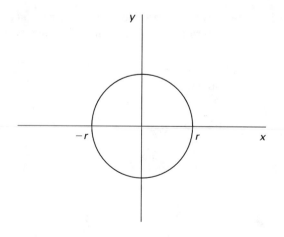

Example 1.40
Find the curvature of $y = x^3$ at the point $(1, 0)$.

Solution:
First find y' and y''.

$$y' = 3x^2$$
$$y'' = 6x$$

From Eq. 1.25,

$$K = \left| \frac{6x}{[1+(3x^2)^2]^{\frac{3}{2}}} \right|$$

At the point $(1, 0)$, $x = 1$ and $y = 0$.

$$K = \left| \frac{(6)(1)}{\left(1+[(3)(1)^2]^2\right)^{\frac{3}{2}}} \right|$$

$$= \left| \frac{6}{10^{\frac{3}{2}}} \right| \sim 0.19$$

PRACTICE PROBLEMS

13. Find the curvature of $y = \ln x$ at the point $(2, \ln 2)$.

14. At which point does the curve $y = x^2$ have maximum curvature?

5 L'Hôpital's Rule

Certain limits that may otherwise be difficult can often be evaluated using a systematic technique known as *l'Hôpital's rule*. The technique itself is based on differentiation and can be applied to a wide range of limits known as *indeterminate forms*. Before presenting l'Hôpital's rule, however, it is necessary to discuss the different types of indeterminate forms and how each type is identified.

Consider the following limit.

$$\lim_{x \to 1} \frac{x^2 - 1}{x - 1}$$

As x approaches 1, both the numerator and the denominator of the expression $(x^2-1)/(x-1)$ approach zero. It is customary to describe such limits as *indeterminate forms of type* $\frac{0}{0}$. The value of such a limit is generally not evident by inspection. In fact, the word *indeterminate* is used to indicate that some further investigation needs to take place before it can be decided whether or not the limit exists.

To investigate the previous limit, factor the numerator and simplify the resulting expression.

$$\lim_{x \to 1} \frac{x^2 - 1}{x - 1} = \lim_{x \to 1} \frac{(x-1)(x+1)}{x - 1}$$
$$= \lim_{x \to 1} (x + 1) = 2$$

However, not all indeterminate forms can be dealt with in this way. For example, consider the following limit.

$$\lim_{x \to 1} \frac{\ln x}{x - 1}$$

Both the numerator and the denominator of the expression $(\ln x)/(x-1)$ approach zero as x approaches 1. The limit is therefore indeterminate of the type $\frac{0}{0}$. In this case, however, $(\ln x)/(x - 1)$ cannot be simplified in the same way as the earlier limit. Consequently, the previous limit remains indeterminate.

Indeterminate forms also occur in the form $\frac{\infty}{\infty}$. For example,

$$\lim_{x \to \infty} \frac{e^x}{x^3}$$

This limit is called an *indeterminate form of the type* $\frac{\infty}{\infty}$ since both the numerator and the denominator approach ∞ (become very large) as x approaches ∞. The same is true of the following limit.

$$\lim_{x \to \infty} \frac{\ln x}{x}$$

Identifying an Indeterminate Form

Consider the following limit.

$$\lim_{x \to a} \frac{f(x)}{g(x)}$$

To identify this limit as an indeterminate form of the type $\frac{0}{0}$ or $\frac{\infty}{\infty}$, either show that as x approaches a, both the numerator $f(x)$ and the denominator $g(x)$ approach 0 or ∞, respectively, or proceed as follows. Substitute $x = a$ into the expression $f(x)/g(x)$ to obtain either of the following expressions. (This method works even if a is replaced by $\pm\infty$.)

$$\frac{f(a)}{g(a)} = \frac{0}{0} \quad \text{or} \quad \frac{f(a)}{g(a)} = \frac{\infty}{\infty}$$

Example 1.41 ....
Show that the following limit is indeterminate of the type $\frac{0}{0}$.

$$\lim_{x \to 1} \left[\frac{x - 1}{\sin (x - 1)} \right]$$

Solution:
Comparing $\lim_{x \to 1} [x - 1/\sin (x - 1)]$ with $\lim_{x \to a} f(x)/g(x)$, we find $f(x) = x - 1$, $g(x) = \sin (x - 1)$, and $a = 1$.

Forming the quotient $f(a)/g(a)$,

$$\frac{f(a)}{g(a)} = \frac{f(1)}{g(1)} = \frac{1 - 1}{\sin 1 - 1} = \frac{0}{0}$$

This limit is indeterminate of the type $\frac{0}{0}$.

Example 1.42 ....
Identify the limit $\lim_{x \to \infty} [(e^x)/(x + 2)]$ as being indeterminate of the type $\frac{\infty}{\infty}$.

Solution:
Comparing with $\lim_{x \to a} [f(x)/g(x)]$, we find $f(x) = e^x$, $g(x) = x + 2$, and $a = \infty$.

Forming the quotient $f(a)/g(a)$ and noting that since both e^x and $x + 2$ approach ∞ as x approaches ∞,

$$\frac{f(a)}{g(a)} = \frac{f(\infty)}{g(\infty)} = \frac{e^\infty}{\infty + 2} = \frac{\infty}{\infty}$$

The limit is indeterminate of the type $\frac{\infty}{\infty}$.

....

L'Hôpital's rule is stated as follows.

Suppose that the limit $\lim_{x \to a} [f(x)/g(x)]$ has been identified as an indeterminate form of the type $\frac{0}{0}$ or $\frac{\infty}{\infty}$. If f and g are differentiable,

$$\lim_{x \to a} \frac{f(x)}{g(x)} = \lim_{x \to a} \frac{f'(x)}{g'(x)} \qquad (1.27)$$

In other words, for an indeterminate form of type $\frac{0}{0}$ or $\frac{\infty}{\infty}$, the limit of the quotient of the functions is equal to the limit of the quotient of their derivatives.

Note the following.

(a) In Eq. 1.27, the expression $x \to a$ can be replaced by $x \to a^+$, $x \to a^-$, $x \to \infty$, or $x \to -\infty$.

(b) If, in Eq. 1.27, f and g can be differentiated more than once, l'Hôpital's rule can be applied repeatedly until a limit is obtained that is not indeterminate. (This requires that before each application of l'Hôpital's rule, the limit under consideration is identified as an indeterminate form of type $\frac{0}{0}$ or $\frac{\infty}{\infty}$.)

Basic properties of the trigonometric, logarithmic, and exponential functions (including their limiting properties) can be found in App. 3.

Example 1.43 ....
Evaluate the following limit.

$$\lim_{x \to 1} \frac{x^2 - 1}{x - 1}$$

Solution:
This limit was introduced as an indeterminate form of the type $\frac{0}{0}$. Apply l'Hôpital's rule.

$$\lim_{x \to 1} \frac{x^2 - 1}{x - 1} = \lim_{x \to 1} \frac{\frac{d}{dx}(x^2 - 1)}{\frac{d}{dx}(x - 1)}$$

$$= \lim_{x \to 1} \frac{2x}{1}$$

This resulting limit is not indeterminate, because as x approaches 1, the numerator of the expression $(2x)/1$ approaches 2 and the denominator stays constant at 1. It follows that

$$\lim_{x \to 1} \frac{x^2 - 1}{x - 1} = \lim_{x \to 1} \frac{2x}{1} = \frac{2}{1} = 2$$

Example 1.44

Evaluate the following limit.

$$\lim_{x \to 1} \frac{\ln x}{x - 1}$$

Solution:

This limit was introduced as an indeterminate form of the type $\frac{0}{0}$. Apply l'Hôpital's rule.

$$\lim_{x \to 1} \frac{\ln x}{x - 1} = \lim_{x \to 1} \frac{\dfrac{d}{dx}(\ln x)}{\dfrac{d}{dx}(x - 1)}$$

$$= \lim_{x \to 1} \frac{\dfrac{1}{x}}{1} = \lim_{x \to 1} \frac{1}{x}$$

This resulting limit is not indeterminate, because as x approaches 1, the numerator of the expression $1/x$ remains constant at 1 and the denominator approaches 1. It follows that

$$\lim_{x \to 1} \frac{\ln x}{x - 1} = \lim_{x \to 1} \frac{1}{x} = \frac{1}{1} = 1$$

Example 1.45

Evaluate the following limit.

$$\lim_{x \to 0} \frac{\sin x}{x}$$

Solution:

The numerator and denominator of the expression $(\sin x)/x$ both approach zero as x approaches zero. The limit is therefore an indeterminate form of type $\frac{0}{0}$, and l'Hôpital's rule applies.

$$\lim_{x \to 0} \frac{\sin x}{x} = \lim_{x \to 0} \frac{\dfrac{d}{dx}(\sin x)}{\dfrac{d}{dx}(x)} = \lim_{x \to 0} \frac{\cos x}{1}$$

This resulting limit is not indeterminate, because as x approaches zero, the numerator of the expression $(\cos x)/1$ approaches 1 while the denominator remains constant at 1. It follows that

$$\lim_{x \to 0} \frac{\sin x}{x} = \lim_{x \to 0} \frac{\cos x}{1} = \frac{1}{1} = 1$$

Example 1.46

Evaluate the following limit.

$$\lim_{x \to 0} \frac{\sin x}{x^2}$$

Solution:

Since both $\sin x$ and x^2 approach zero as x approaches zero, this limit is an indeterminate form of the type $\frac{0}{0}$. L'Hôpital's rule applies.

$$\lim_{x \to 0} \frac{\sin x}{x^2} = \lim_{x \to 0} \frac{\dfrac{d}{dx}(\sin x)}{\dfrac{d}{dx}(x^2)}$$

$$= \lim_{x \to 0} \frac{\cos x}{2x}$$

This resulting limit is not indeterminate. In fact, as x approaches zero, $\cos x$ approaches 1 and $2x$ approaches zero so that the expression $(\cos x)/2x$ approaches $\frac{1}{0}$. It follows that

$$\lim_{x \to 0} \frac{\sin x}{x^2} = \lim_{x \to 0} \frac{\cos x}{2x} = \frac{1}{0} = \infty$$

The limit does not exist.

Example 1.47

Evaluate the following limit.

$$\lim_{x \to \infty} \frac{e^x}{x^3}$$

Solution:

This limit was introduced as an indeterminate form of the type $\frac{\infty}{\infty}$. Apply l'Hôpital's rule.

$$\lim_{x \to \infty} \frac{e^x}{x^3} = \lim_{x \to \infty} \frac{\dfrac{d}{dx}(e^x)}{\dfrac{d}{dx}(x^3)}$$

$$= \lim_{x \to \infty} \frac{e^x}{3x^2}$$

Since both e^x and $3x^2$ approach ∞ as x approaches ∞, this resulting limit remains indeterminate of type $\frac{\infty}{\infty}$. Apply l'Hôpital's rule a second time.

$$\lim_{x \to \infty} \frac{e^x}{x^3} = \lim_{x \to \infty} \frac{e^x}{3x^2}$$

$$= \lim_{x \to \infty} \frac{\dfrac{d}{dx}(e^x)}{\dfrac{d}{dx}(3x^2)}$$

$$= \lim_{x \to \infty} \frac{e^x}{6x}$$

Since both e^x and $6x$ approach ∞ as x approaches ∞, this limit remains indeterminate of type $\frac{\infty}{\infty}$. Apply l'Hôpital's rule a third time.

$$\lim_{x\to\infty} \frac{e^x}{x^3} = \lim_{x\to\infty} \frac{e^x}{3x^2}$$

$$= \lim_{x\to\infty} \frac{e^x}{6x}$$

$$= \lim_{x\to\infty} \frac{\frac{d}{dx}(e^x)}{\frac{d}{dx}(6x)}$$

$$= \lim_{x\to\infty} \frac{e^x}{6}$$

This final limit is not indeterminate, because as x approaches ∞, the numerator of the expression $e^x/6$ approaches ∞ while the denominator remains constant at 6. It follows that

$$\lim_{x\to\infty} \frac{e^x}{x^3} = \lim_{x\to\infty} \frac{e^x}{6} = \frac{\infty}{6} = \infty$$

This means that the limit does not exist.

Example 1.48

Evaluate the following limit.

$$\lim_{x\to\infty} \frac{\ln x}{x}$$

Solution:
This limit was introduced as an indeterminate form of the type $\frac{\infty}{\infty}$. Apply l'Hôpital's rule.

$$\lim_{x\to\infty} \frac{\ln x}{x} = \lim_{x\to\infty} \frac{\frac{1}{x}}{1} = \lim_{x\to\infty} \frac{1}{x}$$

This resulting limit is not indeterminate, because as x approaches ∞, the numerator of the expression $1/x$ remains constant at 1, while the denominator approaches ∞. It follows that

$$\lim_{x\to\infty} \frac{\ln x}{x} = \lim_{x\to\infty} \frac{1}{x} = \frac{1}{\infty} = 0$$

Before applying l'Hôpital's rule, always check that the limit under consideration is indeed indeterminate of type $\frac{0}{0}$ or $\frac{\infty}{\infty}$; otherwise, l'Hôpital's rule will deliver the wrong answer. For example, applying l'Hôpital's rule to the limit $\lim_{x\to 0} \tan x$,

$$\lim_{x\to 0} \tan x = \lim_{x\to 0} \frac{\sin x}{\cos x}$$

$$= \lim_{x\to 0} \frac{\frac{d}{dx}(\sin x)}{\frac{d}{dx}(\cos x)}$$

$$= \lim_{x\to 0} \frac{\cos x}{(-\sin x)}$$

$$= \frac{1}{0} = \infty$$

This wrongly suggests that $\lim_{x\to 0} \tan x = \infty$ (it does not exist). In fact, from App. 3, $\lim_{x\to 0} \tan x = 0$. The contradiction arises from the fact that the original limit $\lim_{x\to 0} \tan x = \lim_{x\to 0} [\sin x/\cos x]$ is not indeterminate. This fact can be seen by substituting $x = 0$ into the expression $\sin x/\cos x$ to obtain $\frac{0}{1}$. In other words, the limit is not of the type to which l'Hôpital's rule can be applied.

The following examples illustrate further the application of l'Hôpital's rule.

Example 1.48

Evaluate the following expression.

$$\lim_{x\to 1} \frac{x^8 - 1}{x^6 - 1}$$

Solution:
Substituting $x = 1$ into the expression $(x^8 - 1)/(x^6 - 1)$ leads to $\frac{0}{0}$. Therefore, $\lim_{x\to 1}[(x^8 - 1)/(x^6 - 1)]$ is indeterminate of type $\frac{0}{0}$. L'Hôpital's rule applies.

$$\lim_{x\to 1} \frac{x^8 - 1}{x^6 - 1} = \lim_{x\to 1} \frac{\frac{d}{dx}(x^8 - 1)}{\frac{d}{dx}(x^6 - 1)}$$

$$= \lim_{x\to 1} \frac{8x^7}{6x^5} = \lim_{x\to 1} \frac{4x^2}{3}$$

This resulting limit is not indeterminate, because as x approaches 1, the numerator of the expression $(4x^2)/3$ approaches 4 while the denominator remains constant at 3. It follows that

$$\lim_{x\to 1} \frac{x^8 - 1}{x^6 - 1} = \lim_{x\to 1} \frac{4x^2}{3} = \frac{4}{3}$$

Example 1.49

Evaluate the following expression.

$$\lim_{x\to -2} \frac{x + 2}{x^2 - 4}$$

Solution:

Substituting $x = -2$ into the expression $(x+2)/(x^2-4)$ leads to $\frac{0}{0}$. Therefore, $\lim_{x \to -2}[(x+2)/(x^2-4)]$ is indeterminate of type $\frac{0}{0}$. Apply l'Hôpital's rule.

$$\lim_{x \to -2} \frac{x+2}{x^2-4} = \lim_{x \to -2} \frac{\frac{d}{dx}(x+2)}{\frac{d}{dx}(x^2-4)}$$

$$= \lim_{x \to -2} \frac{1}{2x}$$

$$= \frac{1}{2(-2)} = -\frac{1}{4}$$

Alternatively, factor the denominator of the expression $(x+2)/(x^2-4)$.

$$\lim_{x \to -2} \frac{x+2}{x^2-4} = \lim_{x \to -2} \frac{x+2}{(x-2)(x+2)}$$

$$= \lim_{x \to -2} \frac{1}{x-2} = -\frac{1}{4}$$

Example 1.50

Evaluate the following expression.

$$\lim_{x \to 0} \frac{\sin x}{e^{\cos x}}$$

Solution:

Substituting $x = 0$ into the expression $(\sin x)/(e^{\cos x})$ leads to $0/(e^{\cos 0}) = 0/e^1$. Therefore, the limit $\lim_{x \to 0}[(\sin x)/(e^{\cos x})]$ is not indeterminate. L'Hôpital's rule does not apply. In fact, l'Hôpital's rule is not necessary, because as x approaches zero, $\sin x$ approaches zero and $e^{\cos x}$ approaches $e^1 = e$.

$$\lim_{x \to 0} \frac{\sin x}{e^{\cos x}} = \frac{0}{e} = 0$$

Example 1.51

Evaluate the following expression.

$$\lim_{x \to \infty} \frac{\ln(1+x)}{x^2}$$

Solution:

Substituting $x = \infty$ into the expression $[\ln(1+x)]/(x^2)$ leads to $\frac{\infty}{\infty}$ since both $\ln(1+x)$ and x^2 approach ∞ as x approaches ∞. It follows that $\lim_{x \to \infty}[(\ln(1+x))/x^2]$ is indeterminate of type $\frac{\infty}{\infty}$. Apply l'Hôpital's rule.

$$\lim_{x \to \infty} \frac{\ln(1+x)}{x^2} = \lim_{x \to \infty} \frac{\frac{d}{dx}(\ln(1+x))}{\frac{d}{dx}(x^2)}$$

$$= \lim_{x \to \infty} \frac{\frac{1}{1+x}}{2x} = \lim_{x \to \infty} \frac{1}{2x(1+x)}$$

As x approaches ∞, the denominator of the following expression approaches ∞ while the numerator remains constant at 1.

$$\frac{1}{2x(1+x)} = \frac{1}{2x+2x^2}$$

The resulting limit is therefore not indeterminate. In fact,

$$\lim_{x \to \infty} \frac{\ln(1+x)}{x^2} = \lim_{x \to \infty} \frac{1}{2x(1+x)} = \frac{1}{\infty} = 0$$

Example 1.52

Evaluate $\lim_{x \to \infty} x^3 e^{-x}$.

Solution:

Write $\lim_{x \to \infty} x^3 e^{-x} = \lim_{x \to \infty}[x^3/e^x]$. Substitute $x = \infty$ into the expression x^3/e^x. Since both x^3 and e^x approach ∞ as x approaches ∞, the limit is indeterminate of the type $\frac{\infty}{\infty}$. Apply l'Hôpital's rule.

$$\lim_{x \to \infty} \frac{x^3}{e^x} = \lim_{x \to \infty} \frac{\frac{d}{dx}(x^3)}{\frac{d}{dx}(e^x)}$$

$$= \lim_{x \to \infty} \frac{3x^2}{e^x}$$

This limit remains indeterminate (of type $\frac{\infty}{\infty}$) since both $3x^2$ and e^x approach ∞ as x approaches ∞. Apply l'Hôpital's rule a second time.

$$\lim_{x \to \infty} \frac{x^3}{e^x} = \lim_{x \to \infty} \frac{3x^2}{e^x} = \lim_{x \to \infty} \frac{\frac{d}{dx}(3x^2)}{\frac{d}{dx}(e^x)}$$

$$= \lim_{x \to \infty} \frac{6x}{e^x}$$

The resulting limit continues to be indeterminate (of type $\frac{\infty}{\infty}$) since both $6x$ and e^x approach ∞ as x approaches ∞. Apply l'Hôpital's rule a third time.

$$\lim_{x \to \infty} \frac{x^3}{e^x} = \lim_{x \to \infty} \frac{3x^2}{e^x}$$

$$= \lim_{x \to \infty} \frac{6x}{e^x}$$

$$= \lim_{x \to \infty} \frac{\frac{d}{dx}(6x)}{\frac{d}{dx}(e^x)}$$

$$= \lim_{x \to \infty} \frac{6}{e^x}$$

This limit is not indeterminate since e^x approaches ∞ with x. That is, the expression $6/e^x$ approaches $6/\infty$ as x approaches ∞. It follows that

$$\lim_{x \to \infty} \frac{x^3}{e^x} = \lim_{x \to \infty} \frac{6}{e^x}$$

$$= \frac{6}{\infty} = 0$$

PRACTICE PROBLEMS

Evaluate the following limits.

15. $\displaystyle\lim_{x \to -1} \frac{x^3 + 4x^2 + 5x + 2}{x^2 + 2x + 1}$

16. $\displaystyle\lim_{x \to 0} \frac{\sqrt{1+x} - 3}{\sqrt{x} - 1} \qquad [x > 0]$

17. $\displaystyle\lim_{x \to \infty} \frac{3x^2}{\pi e^x}$

18. $\displaystyle\lim_{x \to \infty} \frac{\sin\left(\dfrac{2}{x}\right)}{\sin\left(\dfrac{3}{x}\right)}$

19. $\displaystyle\lim_{\theta \to 0} \frac{\sin^2 \theta}{\theta}$

20. $\displaystyle\lim_{x \to \infty} \frac{x^{10} + 1}{x - 3}$

FE-Style Exam Problems

1. What is the first derivative of the following expression?

$$y = e^{-x} \ln 2x \qquad [x > 0]$$

(A) $\quad e^{-x} \left[\dfrac{1}{x} + \ln 2x \right]$

(B) $\quad e^{-x} \left[\dfrac{1}{x} - \ln 2x \right]$

(C) $\quad e^{-x} \left[\dfrac{1}{\ln 2x} + \ln 2x \right]$

(D) $\quad -e^{-x} \ln 2x + \dfrac{e^{-x}}{2x}$

2. Evaluate the following limit.

$$\lim_{x \to 1} \frac{3x^2 + 2x - 5}{x^4 + 3x^2 - 4}$$

(A) 0

(B) $\dfrac{2}{5}$

(C) $\dfrac{4}{5}$

(D) ∞

3. What is the curvature of the curve $y = f(x) = 1/x$ at the point $(1, 1)$?

(A) -2

(B) $-\dfrac{1}{\sqrt{2}}$

(C) 0

(D) $\dfrac{1}{\sqrt{2}}$

4. What are the minimum and maximum values, respectively, of the function $f(t) = t^3 + 3t^2 - 9t + 5$ on the interval $[-6, 2.5]$?

(A) $-49, 32$

(B) $0, 32$

(C) $7, 35$

(D) $11, 49$

5. If $\varphi(x, y, z) = \dfrac{-1}{\sqrt{x^2 + y^2 + z^2}}$, then $\dfrac{\partial^2 \varphi}{\partial x^2} + \dfrac{\partial^2 \varphi}{\partial y^2} + \dfrac{\partial^2 \varphi}{\partial z^2}$ is equal to

(A) 0

(B) $\dfrac{2xyz}{x^2 + y^2 + z^2}$

(C) $\ln \sqrt{x + y + z}$

(D) $2xyz \ln (x^2 + y^2 + z^2)$

6. If x and y are related by the equation $xe^{yx} = 3$ and $x > 0$, what is dy/dx?

(A) $\dfrac{-(y + \ln x) - 1}{x}$

(B) $\dfrac{-(xy - 1)}{y^2}$

(C) $\dfrac{-(xy + 1)}{x^2}$

(D) $\dfrac{3}{y}$

7. If $f(x, y) = x^3 y + \sqrt{y} x^4 + \cos x + \tan y + \sin^6 y$, what is $\partial f / \partial x$?

(A) $x^3 y + \frac{1}{2} y^{-\frac{1}{2}} x^4 + \sin x$

(B) $3x^2 y + 4x^3 \sqrt{y} - \sin x$

(C) $3x^2 y + 4x^3 \sqrt{y} + \sin x$

(D) $x^2 y + 4x^3 \sqrt{y} - \sin x + \sec^2 y$

8. The value of the constant a such that the function $f(x) = ax^2 + 4x + 13$ has a local maximum at $x = 1$ is

(A) -5

(B) -4

(C) -3

(D) -2

9. Evaluate the following limit.

$$\lim_{x \to \infty} \frac{\ln x^{99}}{x^2}$$

(A) -2

(B) -1

(C) 0

(D) 3

10. Which of the following is *not* correct?

(A) If $x^2 + y^2 = 9$, then $\dfrac{dy}{dx} = \dfrac{x}{y}$

(B) $\dfrac{d}{dx}(\sin^2 x) = \sin 2x$

(C) The slope of the tangent line to the circle $x^2 + y^2 = 5$ at the point $(1, 2)$ is $-1/2$.

(D) $\dfrac{d}{dx}\left(\dfrac{x-1}{x}\right) = \dfrac{2}{x^2}$

2

Integral Calculus

Chapter 1 is concerned with differential calculus. Chapter 2 deals with the second main branch of calculus: *integral calculus*. Here, the fundamental, or "base," problem is the *area problem*: Find the area A of the region D that lies under the curve $y = f(x)$ and between $x = a$ and $x = b$.

index, i, takes the values $i = 1, 2, 3, \ldots, n$; and n is a positive integer. In other words,

$$A \sim f(x_1^*)\Delta x_1 + f(x_2^*)\Delta x_2 + \cdots + f(x_n^*)\Delta x_n$$
$$= \sum_{i=1}^{n} f(x_i^*)\Delta x_i \qquad (2.1)$$

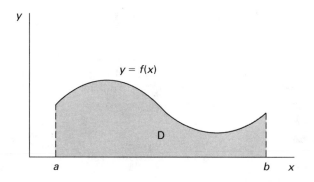

Figure 2.1 Region D

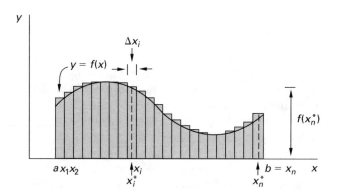

Figure 2.2 Approximating the Area A

The area problem leads to the central concept of integral calculus: the *definite integral*. As in the case of the derivative, the definite integral is defined in terms of limits and is applicable to a wide range of problems from all branches of engineering and applied science.

Differential and integral calculus are connected by an extremely powerful theorem known as the fundamental theorem of calculus, which allows results from differential calculus to be used to solve problems from integral calculus.

1 Fundamental Theorem of Calculus

Consider again the area problem.

From Fig. 2.2, a reasonable approximation to the area, A, is given by the sum of the areas of the individual rectangles of height, $f(x_i^*)$ and width, Δx_i. The point x_i^* can be anywhere in the subinterval $[x_{i-1}, x_i]$; the

Fig. 2.2 indicates that the approximation in Eq. 2.1 will improve as the width, Δx_i, of each rectangle decreases or as the number of rectangles covering the area, A, increases (i.e., as Δx_i approaches zero or as n approaches ∞). Consequently, the area, A, of the region D from the area problem is defined as

$$A = \lim_{n \to \infty} \sum_{i=1}^{n} f(x_i^*)\Delta x_i \qquad (2.2)$$

If the limit in Eq. 2.2 exists, the function, f, is called *integrable* on the interval $[a, b]$, and Eq. 2.2 defines the *definite integral of f from a to b*, denoted by

$$A = \int_a^b f(x)dx = \lim_{n \to \infty} \sum_{i=1}^{n} f(x_i^*)\Delta x_i \qquad (2.3)$$

The symbol $\int$ in Eq. 2.3 is called an *integral sign*. Its "elongated S" appearance reflects the fact that an integral is a limit of sums. In the context of Eq. 2.3, the function $f(x)$ is referred to as the *integrand*, the numbers a and b are called the (lower and upper, respectively) *limits of integration*, and the procedure of calculating an integral is called *integration*.

The definite integral is a number and does not depend on the *variable of integration* x. The variable of integration itself is of little relevance in the evaluation of the definite integral. In fact,

$$\int_a^b f(x)\,dx = \int_a^b f(t)\,dt = \int_a^b f(s)\,ds$$

Consequently, x is referred to as a *dummy variable*.

A definite integral need not represent an area. However, for the special case of $f(x) \geq 0$, as in Eq. 2.1,

$$\int_a^b f(x)\,dx = \text{A} \qquad (2.4)$$

(A is the area under the curve $y = f(x)$ from a to b.)

Evaluating definite integrals from the definition given in Eq. 2.3 involves the calculation of tedious and often difficult limits of sums. As in the case of the derivative, however, it is seldom necessary to resort to the basic definition when evaluating a definite integral. In integral calculus, the fundamental theorem of calculus provides a quick and easy method for evaluating most definite integrals. Before presenting the fundamental theorem, however, it is necessary to examine one of its essential components.

Antiderivatives

Many problems in applied mathematics require the retrieval of a function from its derivative. That is, given the derivative, what is the original function, or *antiderivative*? For example, which function has the derivative $\cos x$? From Table 1.1, one answer is $\sin x$; that is, $\sin x$ is a particular antiderivative of $\cos x$. In fact, there are many other antiderivatives of $\cos x$, but since $(d/dx)(C) = 0$, they all take the form $\sin x + C$, where C is an arbitrary constant.

$$\frac{d}{dx}(\sin x + C) = \frac{d}{dx}(\sin x) + \frac{d}{dx}(C)$$
$$= \cos x + 0$$
$$= \cos x$$

Similarly, from Table 1.1, $F(x) = \ln x$ is a particular antiderivative of $f(x) = 1/x$ where $x > 0$, and $F(x) = e^x$ is a particular antiderivative of $f(x) = e^x$.

For an arbitrary constant C, it follows that $\ln x + C$ is the most general antiderivative of $1/x$ where $x > 0$, and $e^x + C$ is the most general antiderivative of e^x.

In general, if $F(x)$ is a particular antiderivative of some function $f(x)$, the most general antiderivative of $f(x)$ is $F(x) + C$.

The particular antiderivatives of some of the more common functions arising in calculus are listed in the following table.

Table 2.1 Common Antiderivatives

function	particular antiderivative
$x^n \quad [n \neq -1]$	$\dfrac{x^{n+1}}{n+1}$
$\cos x$	$\sin x$
$\sin x$	$-\cos x$
e^x	e^x
$\sec^2 x$	$\tan x$
$c_1 f(x) + c_2 g(x)$	$c_1 F(x) + c_2 G(x)$

In Table 2.1, c_1 and c_2 are constant, and G is a particular antiderivative of g. To obtain the most general antiderivatives from Table 2.1, add an arbitrary constant to each particular antiderivative.

Example 2.1 ....

Find the most general antiderivative of

$$f(x) = 3\cos x - 2x^{\frac{1}{2}} + 2\sin x + 3x^2$$

Solution:

First find the particular antiderivative of $f(x)$ using Table 2.1. Do this by finding the particular antiderivative of each term in $f(x)$.

The particular antiderivative of $3\cos x$ is $(3)(\sin x)$.

The particular antiderivative of $-2x^{\frac{1}{2}}$ is

$$(-2)\left(\frac{x^{\frac{1}{2}+1}}{\frac{1}{2}+1}\right) = (-2)\left(\frac{x^{\frac{3}{2}}}{\frac{3}{2}}\right)$$

The particular antiderivative of $2\sin x$ is $(2)(-\cos x)$.

The particular antiderivative of $3x^2$ is

$$(3)\left(\frac{x^{2+1}}{2+1}\right) = (3)\left(\frac{x^3}{3}\right)$$

When these are added together, the particular antiderivative of $f(x)$ is

$$F(x) = (3)(\sin x) - (2)\left(\frac{x^{\frac{3}{2}}}{\frac{3}{2}}\right)$$

$$+ (2)(-\cos x) + (3)\left(\frac{x^3}{3}\right)$$

$$= (3)(\sin x) - \frac{4}{3}x^{\frac{3}{2}} - 2\cos x + x^3$$

The most general antiderivative is found by adding an arbitrary constant C to the particular antiderivative $F(x)$.

$$F(x) + C = 3\sin x - \frac{4}{3}x^{\frac{3}{2}} - 2\cos x + x^3 + C$$

Note that it is not necessary to add an arbitrary constant for each of the terms in $F(x)$. In fact, if C_1, C_2, C_3, and C_4 are arbitrary constants, then $C = C_1 + C_2 + C_3 + C_4$ is also an arbitrary constant, and

$$F(x) + C_1 + C_2 + C_3 + C_4 = F(x) + C$$

In other words, the final answer remains unchanged.

The concept of an antiderivative is central to the main result of this section.

Fundamental Theorem of Calculus
The definite integral in Eq. 2.3 can be evaluated using antiderivatives as follows.

$$\int_a^b f(x)\,dx = [F(x)]_{x=a}^{x=b} = F(b) - F(a)$$

Here, $F(x)$ is a particular antiderivative of the integrand $f(x)$; that is, $F'(x) = f(x)$.

Example 2.2
Evaluate the following definite integral.

$$\int_{-1}^5 x^5\,dx$$

Solution:
First find a particular antiderivative $F(x)$ of the integrand $f(x) = x^5$. From Table 2.1, $F(x) = x^6/6$. By the fundamental theorem of calculus,

$$\int_{-1}^5 x^5\,dx = F(5) - F(-1)$$

$$= \frac{(5)^6}{6} - \frac{(-1)^6}{6} = 2604$$

Example 2.3
Evaluate the following.

$$\int_0^\pi 3\cos\theta\,d\theta$$

Solution:
From Table 2.1, a particular antiderivative of $f(\theta) = 3\cos\theta$ is $F(\theta) = 3\sin\theta$. By the fundamental theorem of calculus,

$$\int_0^\pi 3\cos\theta\,d\theta = F(\pi) - F(0)$$

$$= (3)(\sin\pi) - (3)(\sin 0)$$

$$= (3)(\sin\pi) - 0$$

$$= (3)(0) = 0$$

[Despite the fact that the integrands in Exs. 2.2 and 2.3 are relatively simple, trying to evaluate either of the corresponding definite integrals using the basic (limit) definition in Eq. 2.3 is tedious and difficult. Even in these simple cases, the fundamental theorem of calculus proves to be extremely useful.]

Example 2.4
Find the area under the curve $y = f(x) = x^3$ between $x = 0$, $x = 2$, and the x-axis.

Solution:
From Eqs. 2.3 and 2.4, since $f(x) = x^3 \geq 0$ for x between 0 and 2, the required area is given by the following definite integral.

$$A = \int_0^2 x^3\,dx$$

By the fundamental theorem of calculus,

$$A = \int_0^2 x^3\,dx = F(2) - F(0)$$

Here, the equation $F(x) = x^4/4$ is a particular antiderivative of the integrand $f(x) = x^3$ (from Table 2.1). The area, A, is now given by

$$A = F(2) - F(0)$$

$$= \frac{(2)^4}{4} - \frac{(0)^4}{4}$$

$$= \frac{16}{4} - 0 = 4$$

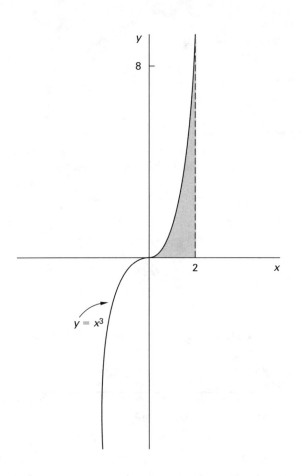

$y = x^3$

Table 2.2 Indefinite Integrals

$$\int k\,dx = kx + C$$

$$\int x^n\,dx = \frac{x^{n+1}}{n+1} + C \quad [n \neq -1]$$

$$\int \frac{1}{x}\,dx = \ln|x| + C$$

$$\int e^{kx}\,dx = \frac{e^{kx}}{k} + C \quad [k \neq 0]$$

$$\int xe^{kx}\,dx = \frac{e^{kx}(kx-1)}{k^2} + C \quad [k \neq 0]$$

$$\int k^{ax}\,dx = \frac{k^{ax}}{a\ln k} + C \quad [k > 0 \text{ and } a \neq 0]$$

$$\int \ln x\,dx = x\ln x - x + C$$

$$\int \sin x\,dx = -\cos x + C$$

$$\int \cos x\,dx = \sin x + C$$

$$\int \tan x\,dx = \ln|\sec x| + C$$

$$\int \cot x\,dx = \ln|\sin x| + C$$

$$\int \sec x\,dx = \ln|\sec x + \tan x| + C$$

$$\int \csc x\,dx = \ln|\csc x - \cot x| + C$$

$$\int \frac{dx}{x^2 + k^2} = \frac{1}{k}\arctan\left(\frac{x}{k}\right) + C \quad [k > 0]$$

$$\int \frac{dx}{\sqrt{k^2 - x^2}} = \arcsin\left(\frac{x}{k}\right) + C \quad [k \neq 0]$$

$$\int \sin^2 x\,dx = \frac{x}{2} - \frac{\sin 2x}{4} + C$$

$$\int \cos^2 x\,dx = \frac{x}{2} + \frac{\sin 2x}{4} + C$$

$$\int \tan^2 x\,dx = \tan x - x + C$$

$$\int kf(x)\,dx = k\int f(x)\,dx$$

$$\int [f(x) + g(x)]\,dx = \int f(x)\,dx + \int g(x)\,dx$$

The strong relationship arising from the fundamental theorem of calculus between antiderivatives and integrals is responsible for the following notation traditionally used for the most general antiderivative of f.

$$\int f(x)\,dx \qquad (2.5)$$

Eq. 2.5 is also referred to as the *indefinite integral of f*. That is, if $F(x)$ is an antiderivative of $f(x)$, it is customary to write:

$$F'(x) = f(x) \text{ or } F(x) = \int f(x)\,dx$$

It is important to distinguish between the two types of integrals: The *indefinite integral* (Eq. 2.5) is a function of x, while the *definite integral* (Eq. 2.3) is a number.

The following table is an expanded version of Table 2.1. Most of the entries come from the table of derivatives, Table 1.1, used in reverse. Others are derived using the techniques presented later in this chapter.

In Table 2.2, k, a, and n are real constants, while C is an arbitrary constant (of integration). Note that no constant of integration appears in the last two formulas since the actual integration has not yet been performed (these are simplifying rules applied before the integration takes place).

In the following examples, C is an arbitrary constant of integration.

Example 2.5

Determine the following indefinite integral.

$$\int \left(\frac{x^3 + x + 2}{\sqrt{x}} \right) dx$$

Solution:

First simplify the integrand.

$$\int \left(\frac{x^3 + x + 2}{\sqrt{x}} \right) dx = \int \left(\frac{x^3}{x^{\frac{1}{2}}} + \frac{x}{x^{\frac{1}{2}}} + \frac{2}{x^{\frac{1}{2}}} \right) dx$$

$$= \int \left(x^{\frac{5}{2}} + x^{\frac{1}{2}} + 2x^{-\frac{1}{2}} \right) dx$$

Next, use Table 2.2.

$$\int \left(\frac{x^3 + x + 2}{\sqrt{x}} \right) dx = \frac{x^{\frac{5}{2}+1}}{\frac{5}{2}+1} + \frac{x^{\frac{1}{2}+1}}{\frac{1}{2}+1}$$

$$+ (2) \left(\frac{x^{-\frac{1}{2}+1}}{-\frac{1}{2}+1} \right) + C$$

$$= \frac{x^{\frac{7}{2}}}{\frac{7}{2}} + \frac{x^{\frac{3}{2}}}{\frac{3}{2}} + (2) \left(\frac{x^{\frac{1}{2}}}{\frac{1}{2}} \right) + C$$

$$= \frac{2}{7}x^{\frac{7}{2}} + \frac{2}{3}x^{\frac{3}{2}} + 4x^{\frac{1}{2}} + C$$

Example 2.6

Determine the following indefinite integral.

$$\int \left(\frac{3}{8 + 6x^2} \right) dx$$

Solution:

The integrand $3/(8 + 6x^2)$ is related to the integrand $1/(x^2 + k^2)$ appearing in Table 2.2. Simplify to identify the number k.

$$\int \left(\frac{3}{8 + 6x^2} \right) dx = 3 \int \left(\frac{1}{(6)\left(\frac{8}{6} + x^2 \right)} \right) dx$$

$$= 3 \int \left(\frac{1}{(6)\left(\frac{4}{3} + x^2 \right)} \right) dx$$

$$= 3 \int \left(\frac{1}{(6)\left[\left(\frac{2}{\sqrt{3}} \right)^2 + x^2 \right]} \right) dx$$

$$= \frac{3}{6} \int \frac{dx}{\left(\frac{2}{\sqrt{3}} \right)^2 + x^2}$$

$$= \frac{1}{2} \int \frac{dx}{\left(\frac{2}{\sqrt{3}} \right)^2 + x^2}$$

The number $k = 2/\sqrt{3}$. From Table 2.2,

$$\int \left(\frac{3}{8 + 6x^2} \right) dx = \frac{1}{2} \int \frac{dx}{\left(\frac{2}{\sqrt{3}} \right)^2 + x^2}$$

$$= \left(\frac{1}{2} \right) \left[\frac{1}{\left(\frac{2}{\sqrt{3}} \right)} \arctan \left(\frac{x}{\frac{2}{\sqrt{3}}} \right) \right] + C$$

$$= \left(\frac{1}{2} \right) \left(\frac{\sqrt{3}}{2} \right) \arctan \left(\frac{\sqrt{3}x}{2} \right) + C$$

$$= \frac{\sqrt{3}}{4} \arctan \left(\frac{\sqrt{3}x}{2} \right) + C$$

Example 2.7

Determine the following indefinite integral.

$$\int \left(e^{3x} + x^2 + \frac{2}{\sqrt{1 - x^2}} + \sin x \right) dx$$

Solution:

Integrate each term in the integrand separately using Table 2.2.

$$\int \left(e^{3x} + x^2 + \frac{2}{\sqrt{1-x^2}} + \sin x \right) dx$$

$$= \int e^{3x}\,dx + \int x^2\,dx + 2\int \frac{dx}{\sqrt{1-x^2}} + \int \sin x\,dx$$

$$= \frac{e^{3x}}{3} + \frac{x^3}{3} + 2\arcsin x - \cos x + C$$

....

Example 2.8

Evaluate the definite integral $\int_0^1 (\sqrt{x} + 1)\,dx$.

Solution:

Apply the fundamental theorem of calculus.

$$\int_0^1 (\sqrt{x} + 1)\,dx = F(1) - F(0)$$

Here $F(x)$ is a particular antiderivative of the function $f(x) = \sqrt{x} + 1$. From Table 2.2,

$$F(x) = \int (\sqrt{x} + 1)\,dx$$

$$= \int (x^{\frac{1}{2}} + 1)\,dx$$

$$= \frac{x^{\frac{3}{2}}}{\frac{3}{2}} + x$$

Finally,

$$\int_0^1 (\sqrt{x} + 1)\,dx = F(1) - F(0)$$

$$= \left(\frac{1^{\frac{3}{2}}}{\frac{3}{2}} + 1 \right) - \left(\frac{0^{\frac{3}{2}}}{\frac{3}{2}} + 0 \right)$$

$$= \left(\frac{2}{3} + 1 \right) - (0) = \frac{5}{3}$$

....

Example 2.9

Evaluate the following definite integral.

$$\int_0^1 \left(\frac{3}{\sqrt{8 - 2x^2}} \right) dx$$

Solution:

The integrand $3/\sqrt{8 - 2x^2}$ is related to the integrand $1/\sqrt{k^2 - x^2}$ from Table 2.2. Simplify to find k.

$$\int_0^1 \left(\frac{3}{\sqrt{8 - 2x^2}} \right) dx = 3\int_0^1 \frac{dx}{\sqrt{(2)(4 - x^2)}}$$

$$= 3\int_0^1 \frac{dx}{(\sqrt{2})(\sqrt{4 - x^2})}$$

$$= \frac{3}{\sqrt{2}} \int_0^1 \frac{dx}{\sqrt{(2)^2 - x^2}}$$

It follows that $k = 2$. From Table 2.2,

$$\int_0^1 \left(\frac{3}{\sqrt{8 - 2x^2}} \right) dx$$

$$= \frac{3}{\sqrt{2}} \int_0^1 \frac{dx}{\sqrt{(2)^2 - x^2}}$$

$$= \left(\frac{3}{\sqrt{2}} \right) \left[\arcsin \left(\frac{x}{2} \right) \right]_0^1$$

$$= \left(\frac{3}{\sqrt{2}} \right) \left[\arcsin \left(\frac{1}{2} \right) - \arcsin \left(\frac{0}{2} \right) \right]$$

$$= \left(\frac{3}{\sqrt{2}} \right) \left(\frac{\pi}{6} - 0 \right) = \frac{\pi}{2\sqrt{2}}$$

....

Example 2.10

Find the area between the lines $y = 2x$, $2y + x - 4 = 0$, and the x-axis as illustrated by the shaded area in the following illustration.

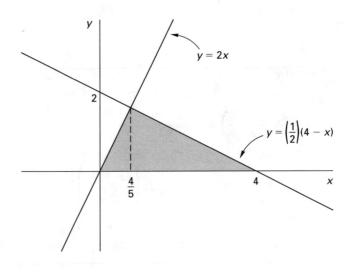

Solution:

From the previous illustration and from Eq. 2.4, the required area, A, is given by the following sum of two definite integrals.

$$A = \int_0^a f_1(x)\,dx + \int_a^b f_2(x)\,dx$$

Here, the functions $f_1(x)$ and $f_2(x)$ correspond to the equations of the given lines written in standard form. That is,

$$y = f_1(x) = 2x \text{ and } y = f_2(x) = \left(\frac{1}{2} \right)(4 - x) \quad \text{(Eq. A)}$$

$x = a$ is the x-coordinate of the point of intersection of the lines in Eq. A, and $x = b$ is the intersection of the line $y = f_2(x) = (1/2)(4 - x)$ with the x-axis. First, locate $x = a$. The lines in Eq. A intersect when

$$f_1(x) = f_2(x)$$

$$2x = \left(\frac{1}{2}\right)(4 - x)$$

$$5x = 4$$

$$x = \frac{4}{5}$$

That is, $x = a = 4/5$. To locate $x = b$, note that the line $y = (1/2)(4 - x)$ intersects the x-axis when

$$y = \left(\frac{1}{2}\right)(4 - x) = 0$$

This equation is satisfied by $x = 4$; that is, $x = b = 4$. The required area is now given by

$$A = \int_0^{\frac{4}{5}} 2x \, dx + \int_{\frac{4}{5}}^4 \left(\frac{1}{2}\right)(4 - x) \, dx$$

Apply the fundamental theorem of calculus to each integral.

$$A = \int_0^{\frac{4}{5}} 2x \, dx + \int_{\frac{4}{5}}^4 \left(\frac{1}{2}\right)(4 - x) dx$$

$$= \left[x^2\right]_0^{\frac{4}{5}} + \left(\frac{1}{2}\right)\left[4x - \frac{x^2}{2}\right]_{\frac{4}{5}}^4$$

$$= \left[\left(\frac{4}{5}\right)^2 - 0^2\right] + \left(\frac{1}{2}\right)$$

$$\times \left(\left[(4)(4) - \frac{4^2}{2}\right] - \left[(4)\left(\frac{4}{5}\right) - \frac{\left(\frac{4}{5}\right)^2}{2}\right]\right)$$

$$= \frac{16}{25} + \left(\frac{1}{2}\right)\left(8 - \frac{72}{25}\right)$$

$$= \frac{16}{5}$$

The required area is $16/5$.

....

The effectiveness of the fundamental theorem of calculus is entirely dependent on the availability of anti-derivatives or indefinite integrals. In Exs. 2.1 through 2.10, Table 2.2 was the main source of indefinite integrals. However, Table 2.2 will not suffice for all cases. For example, the integral $\int 3x^2 \cos(x^3 + 10) \, dx$ does not appear in Table 2.2, or in any standard table of indefinite integrals. To deal with this and other more complicated indefinite integrals, various techniques of (indefinite) integration have been developed. These are examined in detail in the following section.

PRACTICE PROBLEMS

1. Find the area of the region lying beneath the curve $y = x^2 + 1$ from $x = -1$ to $x = 2$ as illustrated in the following figure.

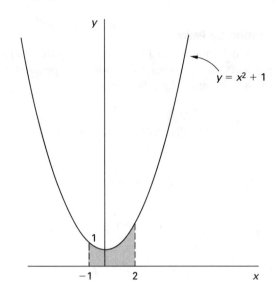

2. Determine the following indefinite integrals.

(a) $\int u(\sqrt{u} + u^3) \, du$

(b) $\int (x + 1)^2 \, dx$

(c) $\int \frac{t^5 - t}{\sqrt{t}} \, dt$

3. Evaluate the following definite integrals using the fundamental theorem of calculus.

(a) $\int_{-\frac{\pi}{2}}^{\frac{\pi}{2}} \cos x \, dx$

(b) $\int_1^3 x^{-3} \, dx$

(c) $\int_{-1}^0 (t - 1)(2t + 2) \, dt$

2 Methods of Integration

There are many important indefinite integrals that cannot be obtained directly from any of the basic formulas given in Table 2.2. Such integrals are referred to as *non-standard integrals*. For example, the indefinite integral $\int x \sin x \, dx$ is nonstandard.

This section is concerned with the development of a series of integration techniques designed to reduce certain nonstandard integrals, such as $\int x \sin x \, dx$, to simpler integrals that can be found from Table 2.2. Following the development of the techniques for differentiation (Eqs. 1.4 through 1.13), integrals will be classified into distinct groups and general strategies and techniques will be developed for each group. This will make the process of integration more systematic.

In what follows, C is an arbitrary constant (of integration).

Integration by Parts

Integration by parts will be used to integrate products or quotients of functions. Consider the product rule (Eq. 1.5) written in the following form.

$$\frac{d}{dx}\left(f(x)g(x)\right) = f(x)g'(x) + g(x)f'(x)$$

Integrate both sides with respect to x.

$$\int \left[\frac{d}{dx}\left(f(x)g(x)\right)\right] dx = \int f(x)g'(x)\,dx$$
$$+ \int g(x)f'(x)\,dx$$
$$f(x)g(x) + C = \int f(x)g'(x)\,dx$$
$$+ \int g(x)f'(x)\,dx$$

Rearrange to obtain the following formula.

$$\int f(x)g'(x)dx = f(x)g(x) - \int g(x)f'(x)dx + C \quad \textbf{(2.6)}$$

Equation 2.6 can be rewritten in the following form.

$$\int f(x)dg(x) = f(x)g(x) - \int g(x)df(x) + C \quad \textbf{(2.7)}$$

In Eq. 2.7, $dg(x) = g'(x)dx$ and $df(x) = f'(x)dx$.

Equation 2.6 leads to a formula for integrating a product of two functions. In fact, replacing $g'(x)$ with some function $h(x)$ (and therefore replacing $g(x)$ with any antiderivative of $h(x)$, for example, $H(x)$) leads to

$$\int f(x)h(x)dx = f(x)H(x) - \int H(x)f'(x)dx + C \quad \textbf{(2.8)}$$

Equation 2.8 is known as *integration by parts*.

Example 2.11
Determine the integral $\int x \sin x\,dx$.

Solution:
The integrand $x \sin x$ is a product of the functions x and $\sin x$. Apply Eq. 2.8 with $f(x) = x$ and $h(x) = \sin x$ so that $H(x) = -\cos x$.

$$\int x \sin x\,dx = x\left(-\cos x\right) - \int (-\cos x)\left(\frac{d}{dx}(x)\right) dx$$
$$= -x \cos x + \int (\cos x)(1)\,dx$$
$$= -x \cos x + \sin x + C$$

(Notice that replacing $H(x)$ by the most general antiderivative of $h(x) = \sin x$, that is, $-\cos x + C$, makes no difference to the final answer.)

The success of integration by parts is entirely dependent on which function is identified as f and which as h in Eq. 2.8. This concept is illustrated in the following example.

If, in Ex. 2.8, the roles of f and h are reversed (i.e., $f(x) = \sin x$ and $h(x) = x$, so that $H(x) = x^2/2$), from Eq. 2.8,

$$\int x \sin x\,dx = (\sin x)\left(\frac{x^2}{2}\right) - \int \frac{x^2}{2} \cos x\,dx + C \quad \textbf{(Eq. A)}$$

The integral $\int (x^2/2)\cos x\,dx$ on the right-hand side of Eq. A is more complicated than the original integral $\int x \sin x\,dx$. This means that the application of integration by parts with this choice of f and h is counterproductive. Integration by parts is only useful if it leads to a simpler integration problem. This means that f and h in Eq. 2.8 must be chosen in such a way that the resulting integral (on the right-hand side of Eq. 2.8) is simpler than the original integral (on the left-hand side of Eq. 2.8). This rule should be used when choosing f and h.

Example 2.12
Determine the indefinite integral $\int xe^x dx$.

Solution:
The integrand xe^x is a product of the two functions x and e^x. Integration by parts (Eq. 2.8) can be used. To determine which of the functions x and e^x is to be identified as $f(x)$ in Eq. 2.8, examine the integral on the right-hand side of Eq. 2.8 for each case. With $f(x) = x$ and $h(x) = e^x$ (so that $f'(x) = 1$ and $H(x) = e^x$), the integral on the right-hand side of Eq. 2.8 becomes

$$\int H(x)f'(x)dx = \int (e^x)(1)dx = \int e^x dx \quad \textbf{(Eq. A)}$$

With $f(x) = e^x$ and $h(x) = x$ (so that $f'(x) = e^x$ and $H(x) = x^2/2$), the integral on the right-hand side of Eq. 2.8 becomes

$$\int H(x)f'(x)dx = \int \left(\frac{x^2}{2}\right)(e^x)dx = \int \left(\frac{x^2}{2}\right)(e^x)dx$$

(Eq. B)

The integral in Eq. A is simpler than the original integral $\int xe^x dx$, whereas the integral in Eq. B is not. Apply Eq. 2.8 with $f(x) = x$, $h(x) = e^x$, and $H(x) = e^x$.

$$\int xe^x dx = x\,(e^x) - \int (e^x)(1)dx$$

$$= xe^x - \int e^x dx$$

$$= xe^x - e^x + C$$

$$= e^x(x - 1) + C$$

Example 2.13

Determine the indefinite integral $\int (x + 2)\cos x\, dx$.

Solution:

The integrand $(x+2)\cos x$ is a product of the two functions $x+2$ and $\cos x$. Integration by parts (Eq. 2.8) can be used. To determine which of the functions $x+2$ and $\cos x$ is to be identified as $f(x)$ in Eq. 2.8, examine the integral on the right-hand side of Eq. 2.8 for each case. With $f(x) = x + 2$ and $h(x) = \cos x$ (so that $f'(x) = 1$ and $H(x) = \sin x$), the integral on the right-hand side of Eq. 2.8 becomes

$$\int H(x)f'(x)dx = \int (\sin x)(1)dx = \int \sin x\, dx \quad \textbf{(Eq. A)}$$

With $f(x) = \cos x$ and $h(x) = x + 2$ (so that $f'(x) = -\sin x$ and $H(x) = x^2/(2 + 2x)$), the integral on the right-hand side of Eq. 2.8 becomes

$$\int H(x)f'(x)dx = \int \left(\frac{x^2}{2} + 2\right)(-\sin x)dx \quad \textbf{(Eq. B)}$$

The integral in Eq. A is simpler than the original integral $\int (x + 2)\cos x\, dx$, whereas the integral in Eq. B is not. Apply Eq. 2.8 with $f(x) = x + 2$, $h(x) = \cos x$, and $H(x) = \sin x$.

$$\int (x + 2)\cos x\, dx = (x + 2)(\sin x) - \int (\sin x)(1)dx$$

$$= (x + 2)(\sin x) - \int \sin x\, dx$$

$$= (x + 2)(\sin x) - (-\cos x) + C$$

$$= (x + 2)(\sin x) + \cos x + C$$

Example 2.14

Determine the indefinite integral $\int x \ln x\, dx$.

Solution:

The integrand $x \ln x$ is a product of the two functions x and $\ln x$. Integration by parts (Eq. 2.8) can be used. To determine which of the functions x and $\ln x$ is to be identified as $f(x)$ in Eq. 2.8, examine the integral on the right-hand side of Eq. 2.8 for each case. With $f(x) = x$ and $h(x) = \ln x$ (so that $f'(x) = 1$ and, from Table 2.2, $H(x) = x \ln x - x$), the integral on the right-hand side of Eq. 2.8 becomes

$$\int H(x)f'(x)dx = \int (x \ln x - x)(1)dx = \int (x \ln x - x)dx$$

(Eq. A)

With $f(x) = \ln x$ and $h(x) = x$ (so that $f'(x) = 1/x$ and $H(x) = x^2/2$), the integral on the right-hand side of Eq. 2.8 becomes

$$\int H(x)f'(x)dx = \int \left(\frac{x^2}{2}\right)\left(\frac{1}{x}\right)dx = \int \frac{x}{2}\, dx \quad \textbf{(Eq. B)}$$

The integral in Eq. B is simpler than the original integral $\int x \ln x\, dx$, whereas the integral in Eq. A is not. Apply Eq. 2.8 with $f(x) = \ln x$, $h(x) = x$, and $H(x) = x^2/2$.

$$\int x \ln x\, dx = \left(\frac{x^2}{2}\right)\ln x - \int \left(\frac{x^2}{2}\right)\left(\frac{1}{x}\right)dx$$

$$= \left(\frac{x^2}{2}\right)\ln x - \int \frac{x}{2}\, dx$$

$$= \left(\frac{x^2}{2}\right)\ln x - \frac{x^2}{4} + C$$

Example 2.15

Determine $\int x^2 \sin x\, dx$.

Solution:

Proceeding as in Exs. 2.12 through 2.14, apply integration by parts (Eq. 2.8) with $f(x) = x^2$, $h(x) = \sin x$, and $H(x) = -\cos x$.

$$\int x^2 \sin x\, dx = x^2(-\cos x) - \int (-\cos x)(2x)\, dx$$

$$= -x^2(\cos x) + 2\int x \cos x\, dx$$

Apply integration by parts (Eq. 2.8) once more, this time to the resulting integral. Choosing $f(x) = x$ and $h(x) = \cos x$,

$$\int x \cos x\, dx = x(\sin x) - \int (\sin x)(1)\, dx$$

$$= x \sin x - \int \sin x\, dx$$

$$= x \sin x - (-\cos x)$$

$$= x \sin x + \cos x + C$$

Finally,

$$\int x^2 \sin x \, dx = -x^2 \cos x + 2 \int x \cos x \, dx$$

$$= -x^2 \cos x + (2)(x \sin x + \cos x) + C$$

Example 2.16

Determine $\int e^x \cos x \, dx$.

Solution:

Apply integration by parts, Eq. 2.8. In this case, the integral on the right-hand side of Eq. 2.8 does not simplify whether one chooses $f(x) = e^x$ or $f(x) = \cos x$. Let $f(x) = e^x$ and apply Eq. 2.8.

$$\int e^x \cos x \, dx = (e^x)(\sin x) - \int (e^x)(\sin x) \, dx$$

Apply integration by parts (Eq. 2.8) again, this time to the resulting integral $\int e^x \sin x \, dx$. Choose $f(x) = e^x$ and $h(x) = \sin x$.

$$\int e^x \sin x \, dx = e^x(-\cos x) - \int e^x(-\cos x) \, dx$$

$$= -e^x \cos x + \int e^x \cos x \, dx$$

Return now to the original integral.

$$\int e^x \cos x \, dx = e^x \sin x - \int e^x \sin x \, dx$$

$$= e^x \sin x - \left(-e^x \cos x + \int e^x \cos x \, dx \right)$$

$$= e^x \sin x + e^x \cos x - \int e^x \cos x \, dx + C$$

At this stage it might seem that no progress is being made. However, writing $I = \int e^x \cos x \, dx$, the last equation becomes

$$I = e^x \sin x + e^x \cos x - I + C$$

Solving for I,

$$I = \left(\frac{1}{2} \right) [e^x(\sin x + \cos x) + C]$$

$$= \frac{1}{2} e^x(\sin x + \cos x) + C_1$$

Here, the arbitrary constant $C_1 = (1/2)C$. Finally,

$$I = \int e^x \cos x \, dx = \frac{1}{2} e^x(\sin x + \cos x) + C_1$$

The fundamental theorem of calculus, applied in conjunction with Eq. 2.8 for integration by parts, leads to the following formula for integration by parts for *definite integrals*.

$$\int_a^b f(x)h(x) \, dx = \left[f(x)H(x) \right]_a^b - \int_a^b H(x)f'(x) \, dx \quad \textit{(2.9)}$$

Equation 2.9 is applied in the same way as Eq. 2.8 except that the answer is now a number rather than a function of x (since the integrals in Eq. 2.9 are definite integrals).

Example 2.17

Evaluate the following definite integral.

$$\int_1^9 \sqrt{x} \ln x \, dx$$

Solution:

Apply Eq. 2.9 with $f(x) = \ln x$ and $h(x) = \sqrt{x}$ (for the same reasons as those mentioned in Exs. 2.12 through 2.14).

$$\int_1^9 \sqrt{x} \ln x \, dx$$

$$= \left[(\ln x) \left(\frac{x^{\frac{3}{2}}}{\frac{3}{2}} \right) \right]_1^9 - \int_1^9 \left(\frac{x^{\frac{3}{2}}}{\frac{3}{2}} \right) \left(\frac{1}{x} \right) dx$$

$$= \left(\frac{2}{3} \right) \left[(\ln x)(x^{\frac{3}{2}}) \right]_1^9 - \frac{2}{3} \int_1^9 (x^{\frac{3}{2}}) \left(\frac{1}{x} \right) dx$$

$$= \left(\frac{2}{3} \right) \left[(\ln 9)9^{\frac{3}{2}} - (\ln 1)1^{\frac{3}{2}} \right] - \frac{2}{3} \int_1^9 x^{\frac{1}{2}} \, dx$$

$$= \left(\frac{2}{3} \right)(27 \ln 9) - \left(\frac{2}{3} \right)(0) - \left(\frac{2}{3} \right) \left[\frac{x^{\frac{3}{2}}}{\frac{3}{2}} \right]_1^9$$

$$= \left(\frac{2}{3} \right) \left(27 \ln 9 - \left(\frac{2}{3} \right) \left[x^{\frac{3}{2}} \right]_1^9 \right)$$

$$= \left(\frac{2}{3} \right) \left[27 \ln 9 - \left(\frac{2}{3} \right)(9^{\frac{3}{2}} - 1^{\frac{3}{2}}) \right]$$

$$= \left(\frac{2}{3} \right) \left[27 \ln 9 - \left(\frac{2}{3} \right)(27 - 1) \right]$$

$$= \left(\frac{2}{3} \right) \left(27 \ln 9 - \frac{52}{3} \right) \sim 28.0$$

Example 2.18

Evaluate the following definite integral.

$$\int_1^e \sqrt{x} \ln x^2 \, dx$$

Solution:

First note that $\ln x^2 = 2\ln x$. Therefore,

$$\int_1^e \sqrt{x}\ln x^2 dx = 2\int_1^e \sqrt{x}\ln x\, dx$$

Apply Eq. 2.9 with $f(x) = \ln x$ and $h(x) = \sqrt{x} = x^{\frac{1}{2}}$.

$$2\int_1^e x^{\frac{1}{2}}\ln x\, dx$$

$$= (2)\left(\left[(\ln x)\left(\frac{x^{\frac{3}{2}}}{\frac{3}{2}}\right)\right]_1^e - \int_1^e \left(\frac{x^{\frac{3}{2}}}{\frac{3}{2}}\right)\left(\frac{1}{x}\right)dx\right)$$

$$= (2)\left(\frac{2}{3}\right)\left[(\ln e)e^{\frac{3}{2}} - (\ln 1)(1)^{\frac{3}{2}}\right]$$

$$\quad - (2)\left(\frac{2}{3}\right)\int_1^e \frac{x^{\frac{3}{2}}}{x}dx$$

$$= (2)\left(\frac{2}{3}\right)\left[(\ln e)e^{\frac{3}{2}} - (\ln 1)(1) - \int_1^e x^{\frac{1}{2}}dx\right]$$

$$= (2)\left(\frac{2}{3}\right)\left[(1)e^{\frac{3}{2}} - (0)(1) - \left(\frac{2}{3}\right)\left[x^{\frac{3}{2}}\right]_1^e\right]$$

$$= (2)\left(\frac{2}{3}\right)\left[e^{\frac{3}{2}} - \left(\frac{2}{3}\right)\left(e^{\frac{3}{2}} - 1^{\frac{3}{2}}\right)\right]$$

$$= (2)\left(\frac{2}{3}\right)\left(\frac{e^{\frac{3}{2}}}{3} + \frac{2}{3}\right)$$

$$= \left(\frac{4}{9}\right)\left(e^{\frac{3}{2}} + 2\right)$$

$$\sim \left(\frac{4}{9}\right)\left[(2.72)^{\frac{3}{2}} + 2\right] \sim 2.88$$

Integration by Substitution

This technique can be thought of as the analogue of the chain rule (Eq. 1.8) for differentiation. Consider the integral $\int \sqrt{2x+1}\,dx$. Let $u = 2x+1$ and rewrite the integral in terms of u. The integrand becomes $\sqrt{u}$. Since $du/dx = 2$, the differential dx is related to the differential du by $dx = (1/2)\,du$. Consequently,

$$\int \sqrt{2x+1}\,dx = \int (\sqrt{u})\left(\frac{1}{2}du\right)$$

$$= \frac{1}{2}\int u^{\frac{1}{2}}du$$

$$= \left(\frac{1}{2}\right)\left(\frac{u^{\frac{3}{2}}}{\frac{3}{2}}\right) + C$$

$$= \frac{1}{3}u^{\frac{3}{2}} + C$$

To get the answer in terms of the original variable x, let $u = 2x+1$.

$$\int \sqrt{2x+1}\,dx = \frac{1}{3}u^{\frac{3}{2}} + C$$

$$= \left(\frac{1}{3}\right)(2x+1)^{\frac{3}{2}} + C$$

This technique is known as the *substitution rule*. The idea is to replace a complicated integral by a relatively simple integral by substituting the original variable, x, for a new variable, u. Of course, the most difficult (and most important) part of the substitution rule is making the correct choice of substitution, that is, one that simplifies the integral.

There is no general rule for choosing the correct substitution. However, there are certain cases when a substitution is suggested by the particular form of the integral in question. For example, consider an integral of the following form.

$$\int f\big(g(x)\big)g'(x)dx \qquad (2.10)$$

Make the substitution $u = g(x)$. Since $du/dx = g'(x)$, the differential dx is related to the differential du by $du = g'(x)dx$, or $dx = du/g'(x)$. The previous integral now becomes

$$\int f(u)du \qquad (2.11)$$

In many cases, the integral in Eq. 2.11 is simpler to integrate than the original integral in Eq. 2.10. Once the (simplified) integral in Eq. 2.11 is determined (in terms of the variable u), the substitution $u = g(x)$ is used to express the answer in terms of the original variable, x.

To identify an integral as being of the form of Eq. 2.10, look for a function $g(x)$ whose differential $g'(x)dx$ appears in the integral. The function $g(x)$ need not appear alone; it may be part of a composite function $f\big(g(x)\big)$ as in Eq. 2.10. However, the differential $g'(x)dx$ must appear alone and not as part of another function. Once $g(x)$ is identified, the correct substitution is $u = g(x)$.

Suppose, in the integral under consideration, a function $g(x)$ is identified and its differential $g'(x)dx$, except for a constant factor, appears in the integral. The correct substitution remains $u = g(x)$. For example, in the integral $\int x^3\sqrt{1+x^4}\,dx = \int \sqrt{1+x^4}(x^3dx)$, if $u = g(x) = 1 + x^4$, the differential du is given by $du = g'(x)dx = 4x^3dx$. This differential appears in the integral except for the constant factor 4 (i.e., x^3dx appears in the integral). The choice of substitution, however, remains $u = g(x) = 1 + x^4$. (This choice is explained by noting that the integral can be rewritten in the form $(1/4) \int 4x^3 \sqrt{1 + x^4} \, dx$. This integral is now in the form given in Eq. 2.10, so the substitution $u = g(x) = 1 + x^4$ applies.)

If the integral under consideration is not of the form given in Eq. 2.10, choose the substitution $u = g(x)$ where $g(x)$ is some complicated part of the integral. Again, there are no guarantees, but in many cases, the integral does simplify considerably. The following examples illustrate this procedure.

Example 2.19

Determine the following integral.

$$\int 3x^2 \sqrt{1 + x^3} \, dx$$

Solution:

$$\int 3x^2 \sqrt{1 + x^3} \, dx = \int \sqrt{1 + x^3} \left(3x^2 dx\right)$$

Since its differential $du = g'(x)dx = 3x^2 dx$ appears in the integral, make the substitution $u = g(x) = 1 + x^3$. (The integral is therefore of the form given in Eq. 2.10.) Then, $du/dx = 3x^2$ and $du = 3x^2 dx$ or $dx = du/3x^2$. In terms of the new variable, u, the integral becomes

$$\int 3x^2 \sqrt{1 + x^3} \, dx = \int \sqrt{u} \, du$$

$$= \frac{2}{3} u^{\frac{3}{2}} + C$$

Finally, change back to the original variable, x, by substituting $u = 1 + x^3$.

$$\int 3x^2 \sqrt{1 + x^3} \, dx = \frac{2}{3} u^{\frac{3}{2}} + C$$

$$= \left(\frac{2}{3}\right) \left(1 + x^3\right)^{\frac{3}{2}} + C$$

Example 2.20

Determine the integral $\int \sin(ax + b)dx$, where a and b are constant.

Solution:

Make the substitution $u = g(x) = ax + b$ since the differential is $du = g'(x)dx = a \, dx$, which, apart from the constant factor a, appears in the integral (i.e., dx appears in the integral). Then $u = ax + b$, $du/dx = a$, and $dx = du/a$. In terms of the new variable, u, the integral becomes

$$\int \sin(ax + b) \, dx = \int \sin u \, \frac{du}{a}$$

$$= \frac{1}{a} \int \sin u \, du$$

$$= -\frac{1}{a} \cos u + C$$

Finally, return to the original variable, x.

$$\int \sin(ax + b) \, dx = -\frac{1}{a} \cos(ax + b) + C$$

Example 2.21

Determine the following integral.

$$\int \left(\frac{4x^3}{\sqrt{1 + x^4}}\right) dx$$

Solution:

Make the substitution $u = g(x) = 1 + x^4$ since its differential $du = g'(x)dx = 4x^3 dx$ appears in the integral. (The integral is therefore of the form given in Eq. 2.10.) Then $du/dx = 4x^3$ so that $du = 4x^3 dx$. In terms of the new variable, u, the integral becomes

$$\int \frac{4x^3 dx}{\sqrt{1 + x^4}} = \int \frac{du}{\sqrt{u}}$$

$$= \int u^{-\frac{1}{2}} du$$

$$= 2u^{\frac{1}{2}} + C$$

Finally, let $u = 1 + x^4$ and return to the original variable, x.

$$\int \left(\frac{4x^3}{\sqrt{1 + x^4}}\right) dx = 2\sqrt{1 + x^4} + C$$

Example 2.22

Determine the integral $\int (ax + b)^n dx$ where a, b, and n are constant.

Solution:

Using the same reasoning as in Ex. 2.20, make the substitution $u = ax + b$ so that $du/dx = a$ and $du = a \, dx$. In terms of the new variable, u, the integral becomes

$$\int (ax + b)^n dx = \int u^n \frac{du}{a} = \frac{1}{a} \int u^n du$$

According to Table 2.2, this integral has different values depending on the value of n. In fact, from Table 2.2,

$$\int (ax + b)^n dx = \begin{cases} \left(\dfrac{1}{a}\right)\left(\dfrac{u^{n+1}}{n+1}\right) + C, & n \neq -1 \\ \dfrac{1}{a} \ln|u| + C, & n = -1 \end{cases}$$

To return to the original variable, x, let $u = ax + b$.

$$\int (ax + b)^n dx = \begin{cases} \left(\dfrac{1}{a}\right)\left[\dfrac{(ax+b)^{n+1}}{n+1}\right] + C, & n \neq -1 \\ \dfrac{1}{a} \ln|ax + b| + C, & n = -1 \end{cases}$$

Example 2.23..

Determine the integral $\int \tan x \, dx$.

Solution:

First note that

$$\int \tan x \, dx = \int \frac{\sin x}{\cos x} \, dx = \int \frac{(\sin x) dx}{\cos x}$$

Make the substitution $u = g(x) = \cos x$ since its differential is $du = g'(x) dx = -(\sin x) dx$, which, apart from the constant factor -1, appears in the integral. (Note that $u = g(x) = \sin x$ is not a good choice of substitution since its differential is $du = g'(x) dx = (\cos x) dx$, which does not appear in the integral.) Then $du = -(\sin x) dx$, and the integral becomes

$$\int \tan x \, dx = \int \frac{(\sin x) dx}{\cos x}$$
$$= \int \frac{-du}{u}$$
$$= -\ln |u| + C$$
$$= \ln |u|^{-1} + C$$
$$= \ln \frac{1}{|u|} + C$$

Return to the original variable x by substituting $u = \cos x$.

$$\int \tan x \, dx = \ln \frac{1}{|\cos x|} + C = \ln |\sec x| + C$$

This equation agrees with the result in Table 2.2.

....

Example 2.24..

Determine the following integral.

$$\int \frac{1}{x\sqrt{\ln x}} \, dx$$

Solution:

Write the integral in the following form.

$$\int \left(\frac{1}{\sqrt{\ln x}} \right) \left(\frac{1}{x} \right) dx$$

Make the substitution $u = \ln x$ since its differential $du = (1/x) dx$ appears in the integral (the integral is of the form of Eq. 2.10). The integral becomes

$$\int \left(\frac{1}{x\sqrt{\ln x}} \right) dx = \int \frac{du}{\sqrt{u}}$$
$$= \int u^{-\frac{1}{2}} du$$
$$= \frac{u^{\frac{1}{2}}}{\frac{1}{2}} + C$$
$$= 2\sqrt{u} + C$$

Finally, let $u = \ln x$ to get the answer in terms of the original variable, x.

$$\int \left(\frac{1}{x\sqrt{\ln x}} \right) dx = 2\sqrt{\ln x} + C$$

....

The following additional integration formulas are proved using the substitution rule as in Exs. 2.21 through 2.24. In the first three formulas, the correct substitution is $u = ax + b$. In the case of the last formula, let $u = h(x)$.

Table 2.3 Indefinite Integrals from the Substitution Rule

$$\int (ax+b)^n dx = \begin{cases} \left(\dfrac{1}{a} \right) \left[\dfrac{(ax+b)^{n+1}}{n+1} \right] + C, \ n \neq -1 \\ \dfrac{1}{a} \ln |ax+b| + C, \qquad\quad n = -1 \end{cases}$$

$$\int \sin (ax+b) \, dx = -\frac{1}{a} \cos (ax+b) + C$$

$$\int \cos (ax+b) \, dx = \frac{1}{a} \sin (ax+b) + C$$

$$\int \left(\frac{h'(x)}{h(x)} \right) dx = \ln |h(x)| + C$$

Example 2.25..

Determine the following indefinite integral.

$$\int \sin^2 x \cos x \, dx$$

Solution:

Make the substitution $u = \sin x$ since its differential $du = \cos x \, dx$ appears in the integral. In terms of the new variable, u, the integral becomes

$$\int \sin^2 x \cos x \, dx = \int u^2 du$$
$$= \frac{u^3}{3} + C$$

Finally, to return to the original variable, x, let $u = \sin x$.

$$\int \sin^2 x \cos x \, dx = \frac{u^3}{3} + C$$
$$= \frac{\sin^3 x}{3} + C$$

....

Example 2.26..

Determine the following indefinite integral.

$$\int \frac{4x^3 + 6x^2 + 6x + 101}{x^4 + 2x^3 + 3x^2 + 101x - 11} \, dx$$

Solution:
Notice that the numerator of the integrand is exactly the derivative of the denominator. From the last formula in Table 2.3, with $h(x) = x^4 + 2x^3 + 3x^2 + 101x - 11$,

$$\int \left(\frac{4x^3 + 6x^2 + 6x + 101}{x^4 + 2x^3 + 3x^2 + 101x - 11} \right) dx$$
$$= \ln |x^4 + 2x^3 + 3x^2 + 101x - 11| + C$$

Alternatively, make the substitution $u = x^4 + 2x^3 + 3x^2 + 101x - 11$ since its differential, $du = (4x^3 + 6x^2 + 6x + 101)\, dx$, appears in the integral.

In terms of the new variable, u, the integral becomes

$$\int \frac{1}{u} \, du = \ln |u| + C$$

Finally, let $u = x^4 + 2x^3 + 3x^2 + 101x - 11$.

$$\int \left(\frac{4x^3 + 6x^2 + 6x + 101}{x^4 + 2x^3 + 3x^2 + 101x - 11} \right) du$$
$$= \ln |x^4 + 2x^3 + 3x^2 + 101x - 11| + C$$

When dealing with definite integrals, the substitution method works in exactly the same way except that, in keeping with the philosophy that everything in the integral must be written in terms of the new variable, u, the *limits of integration* must also be changed.

Example 2.27
Evaluate the following integral.

$$\int_1^2 \frac{1}{(1 + 2x)^2} \, dx$$

Solution:
Make the substitution $u = 1 + 2x$ since, apart from the constant factor 2, its differential, $du = 2\, dx$, appears in the integral (i.e., dx appears in the integral). The limits of integration given in the integral are values of the variable x. They must be changed to their corresponding "u-values." Do this using the substitution $u = 1 + 2x$: When $x = 1$, $u = 1 + 2(1) = 3$; when $x = 2$, $u = 1 + 2(2) = 5$.

In terms of the new variable, u, the integral becomes

$$\int_{x=1}^{x=2} \left(\frac{1}{(1+2x)^2} \right) dx = \frac{1}{2} \int_{u=3}^{u=5} u^{-2} du$$
$$= \left(\frac{1}{2} \right) \left[-u^{-1} \right]_{u=3}^{u=5}$$
$$= -\left(\frac{1}{2} \right) \left(\frac{1}{5} - \frac{1}{3} \right) = \frac{1}{15}$$

Note that when evaluating definite integrals, there is no need to return to the original variable, x. The result is a number that is independent of whether the integral is expressed in terms of x or u.

Example 2.28
Evaluate the following integral.

$$\int_{\frac{\pi}{6}}^{\frac{\pi}{2}} \frac{\cos \theta}{\sin^3 \theta} \, d\theta$$

Solution:
Make the substitution $u = \sin \theta$ since its differential $du = \cos \theta \, d\theta$ appears in the integral (of the form of Eq. 2.10). Change the limits of integration: When $\theta = \pi/6$, $u = \sin \pi/6 = 1/2$; when $\theta = \pi/2$, $u = \sin \pi/2 = 1$.

In terms of the new variable, u, the integral becomes

$$\int_{\theta=\frac{\pi}{6}}^{\theta=\frac{\pi}{2}} \left(\frac{\cos \theta}{\sin^3 \theta} \right) d\theta = \int_{u=\frac{1}{2}}^{u=1} u^{-3} du$$
$$= \left(-\frac{1}{2} \right) \left[u^{-2} \right]_{u=\frac{1}{2}}^{u=1}$$
$$= \left(-\frac{1}{2} \right) (1 - 4)$$
$$= \frac{3}{2}$$

Trigonometric Substitutions
The choice of substitution provides the greatest challenge in the application of the substitution rule. However, just as the integrand in Eq. 2.10 provides clues to the choice of substitution, there are other special classes of integrals in which the integrands suggest appropriate substitutions.

Consider the class of integrals whose integrands contain terms such as $\sqrt{a^2 - x^2}$, $\sqrt{a^2 + x^2}$, and $\sqrt{x^2 - a^2}$, where a is constant. The following trigonometric identities are given in App. 3.

$$\sin^2 x + \cos^2 x = 1$$
$$\sec^2 x = 1 + \tan^2 x$$

These identities are used in conjunction with the following *trigonometric substitutions* to simplify integrals whose integrands contain any of the terms $\sqrt{a^2 - x^2}$, $\sqrt{a^2 + x^2}$, or $\sqrt{x^2 - a^2}$.

- For an integrand containing the term $\sqrt{a^2 - x^2}$, let $x = a \sin \theta$ *(2.12)*
- For an integrand containing the term $\sqrt{a^2 + x^2}$, let $x = a \tan \theta$ *(2.13)*
- For an integrand containing the term $\sqrt{x^2 - a^2}$, let $x = a \sec \theta$ *(2.14)*

Example 2.29

Simplify the term $\sqrt{a^2 - x^2}$ using the substitution suggested in Eq. 2.12.

Solution:

If $x = a\sin\theta$,

$$a^2 - x^2 = a^2 - a^2\sin^2\theta$$
$$= a^2(1 - \sin^2\theta)$$
$$= a^2\cos^2\theta$$

Consequently,

$$\sqrt{a^2 - x^2} = \sqrt{a^2\cos^2\theta} = a\cos\theta$$

In an integration problem, the term $a\cos\theta$ is more convenient to work with than the term $\sqrt{a^2 - x^2}$.

Example 2.30

Determine the following integral.

$$\int \frac{x^2}{\sqrt{4 - x^2}}\, dx$$

Solution:

Equation 2.12 suggests the substitution $x = 2\sin\theta$. The differential is $dx = 2\cos\theta\, d\theta$, and the integral becomes

$$\int \left(\frac{x^2}{\sqrt{4 - x^2}}\right) dx = \int \frac{4\sin^2\theta}{\sqrt{4 - 4\sin^2\theta}}(2\cos\theta)d\theta$$

$$= \int \frac{4\sin^2\theta}{\sqrt{(4)(1 - \sin^2\theta)}}(2\cos\theta)d\theta$$

$$= \int \frac{4\sin^2\theta}{\sqrt{4\cos^2\theta}}(2\cos\theta)d\theta$$

$$= \int \frac{4\sin^2\theta}{2\cos\theta}(2\cos\theta)d\theta$$

$$= 4\int \sin^2\theta\, d\theta$$

From Table 2.2,

$$\int \left(\frac{x^2}{\sqrt{4 - x^2}}\right) dx = (4)\left(\frac{\theta}{2} - \frac{\sin 2\theta}{4}\right) + C$$

$$= 2\theta - \sin 2\theta + C$$

To return to the original x-variable, note that the substitution $x = 2\sin\theta$ implies that $\sin\theta = x/2$. From the following figure,

$$\cos\theta = \frac{\sqrt{4 - x^2}}{2}$$

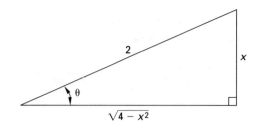

Consequently,

$$\sin 2\theta = 2\sin\theta\cos\theta$$

$$= (2)\left(\frac{x}{2}\right)\left(\frac{\sqrt{4 - x^2}}{2}\right)$$

$$= \frac{x\sqrt{4 - x^2}}{2}$$

Finally,

$$\int \left(\frac{x^2}{\sqrt{4 - x^2}}\right) dx = 2\theta - \sin 2\theta + C$$

$$= 2\arcsin\left(\frac{x}{2}\right) - \frac{x\sqrt{4 - x^2}}{2} + C$$

Other trigonometric integrals are dealt with using the same procedure as in Ex. 2.30 and the substitutions in Eqs. 2.12 through 2.14.

Integration Using Partial Fractions

In elementary algebra, fractions are combined by taking a common denominator. For example,

$$\frac{2}{x + 1} + \frac{3}{x - 2} = \frac{(2)(x - 2) + (3)(x + 1)}{(x + 1)(x - 2)} = \frac{5x - 1}{x^2 - x - 2}$$

When integrating an expression such as that appearing on the far right-hand side of the previous equation, it is useful to have a method that reverses the process of taking a common denominator, that is, one that decomposes the right-hand side of the equation into the simpler (for integration purposes) left-hand side. Such a method exists and is commonly known as the *method of partial fractions*. The procedure is illustrated in the following example.

Consider the expression $(5x - 1)/(x^2 - x - 2)$ and write

$$\frac{5x - 1}{x^2 - x - 2} = \frac{5x - 1}{(x + 1)(x - 2)} = \frac{A}{x + 1} + \frac{B}{x - 2}$$

The terms on the far right-hand side of the previous equation are referred to as *partial fractions*. The constants A and B are determined as follows. First, *clear the fractions* by multiplying through the previous equation by the denominator $(x+1)(x-2)$.

$$\frac{5x-1}{(x+1)(x-2)} = \frac{A}{x+1} + \frac{B}{x-2}$$

$$5x-1 = A(x-2) + B(x+1)$$

Collect terms with the same power of x.

$$5x-1 = (A+B)x + (B-2A)$$

Next, *equate coefficients* of the different powers of x.

coefficient of $x^1 = x$: $\qquad\qquad 5 = A+B$
coefficient of $x^0 = 1$ (constant term): $\quad -1 = B - 2A$

Solving these two algebraic equations using Cramer's rule or elimination leads to $A=2$ and $B=3$. It follows that

$$\frac{5x-1}{x^2-x-2} = \frac{5x-1}{(x+1)(x-2)}$$

$$= \frac{2}{x+1} + \frac{3}{x-2}$$

Note that there is a quicker way to find the constants A and B. Recall the following equation obtained above.

$$5x-1 = A(x-2) + B(x+1)$$

This equation is an identity and holds for every value of x. Choosing specific values of x leads to different simplified forms of the equation from which it is easier to solve for A and B. For example, choosing x such that $x-2=0$ or $x=2$ leads to

$$(5)(2) - 1 = A(2-2) + B(2+1)$$
$$9 = 3B$$
$$B = 3$$

Similarly, choosing $x+1=0$ or $x=-1$ leads to

$$(5)(-1) - 1 = A(-1-2) + B(-1+1)$$
$$-6 = -3A$$
$$A = 2$$

In general, if $P(x)/Q(x)$ is a *proper rational function,* that is, if P and Q are polynomials such that the *degree* of P is less than that of Q, then $P(x)/Q(x)$ decomposes into partial fractions as follows.

$$\frac{P(x)}{Q(x)} = \frac{P(x)}{(a_1x+b_1)(a_2x+b_2) + \cdots + (a_kx+b_k)}$$

$$= \frac{A_1}{a_1x+b_1} + \frac{A_2}{a_2x+b_2} + \cdots + \frac{A_k}{a_kx+b_k} \qquad \text{(2.15)}$$

The constants $A_1, \ldots, A_k$ are determined as in the previous example.

Example 2.31 ....
Determine the following integral.

$$\int \frac{x}{x^3 - 6x^2 + 11x - 6}\, dx$$

Solution:
Identify the integrand with the form $P(x)/Q(x)$. Here, $P(x) = x$ (polynomial of degree 1) and $Q(x) = x^3 - 6x^2 + 11x - 6$ (polynomial of degree 3). The integrand is therefore a proper rational function. To obtain partial fractions for the integrand, first factor the denominator. The expression $x^3 - 6x^2 + 11x - 6$ can be factored into a product of three factors.

$$x^3 - 6x^2 + 11x - 6 = (x-1)(x-2)(x-3)$$

According to Eq. 2.15, the integrand has the following partial fractions.

$$\frac{x}{x^3 - 6x^2 + 11x - 6} = \frac{x}{(x-1)(x-2)(x-3)}$$

$$= \frac{A_1}{(x-1)} + \frac{A_2}{(x-2)} + \frac{A_3}{(x-3)}$$

Next, clear the fractions by multiplying both sides of the equation by the denominator $(x-1)(x-2)(x-3)$.

$$x = A_1(x-2)(x-3) + A_2(x-1)(x-3)$$
$$\quad + A_3(x-1)(x-2)$$
$$= A_1(x^2 - 5x + 6) + A_2(x^2 - 4x + 3)$$
$$\quad + A_3(x^2 - 3x + 2)$$

Collect terms with the same power of x.

$$x = x^2(A_1 + A_2 + A_3) + x(-5A_1 - 4A_2 - 3A_3) + 6A_1$$
$$\quad + 3A_2 + 2A_3$$

Equate coefficients of different powers of x.

$$x^2: 0 = A_1 + A_2 + A_3$$
$$x^1: 1 = -5A_1 - 4A_2 - 3A_3$$
$$x^0: 0 = 6A_1 + 3A_2 + 2A_3$$

Solve these three algebraic equations using Cramer's rule or elimination to obtain the following.

$$A_1 = \frac{1}{2}$$
$$A_2 = -2$$
$$A_3 = \frac{3}{2}$$

Alternatively, substitute, in turn, $x = 1$, $x = 2$, and $x = 3$ into the equation.

$$x = A_1(x-2)(x-3) + A_2(x-1)(x-3) + A_3(x-1)(x-2)$$

Respectively, this leads to

$$1 = 2A_1$$
$$2 = -A_2$$
$$3 = 2A_3$$

The constants are again given by $A_1 = 1/2$, $A_2 = -2$, and $A_3 = 3/2$.

The integral can now be written as

$$\int \frac{x \, dx}{x^3 - 6x^2 + 11x - 6}$$
$$= \int \frac{x \, dx}{(x-1)(x-2)(x-3)}$$
$$= \int \left[\frac{\frac{1}{2}}{(x-1)} - \frac{2}{(x-2)} + \frac{\frac{3}{2}}{(x-3)} \right] dx$$

Finally, from the last formula in Table 2.3,

$$\int \frac{x \, dx}{x^3 - 6x^2 + 11x - 6} = \frac{1}{2}\ln|x-1| - 2\ln|x-2| + \frac{3}{2}\ln|x-3| + C$$

If one of the factors, for example, $(a_1 x + b_1)$, in the denominator Q of the proper rational function P/Q is repeated r times (i.e., the factor $(a_1 x + b_1)^r$ appears in the factored form of $Q(x)$), then instead of the single partial fraction $A_1/(a_1 x + b_1)$, assign r partial fractions as follows.

$$\frac{P(x)}{(a_1 x + b_1)^r} = \frac{A_1}{a_1 x + b_1} + \frac{A_2}{(a_2 x + b_2)^2}$$
$$+ \cdots + \frac{A_r}{(a_r x + b_r)^r} \qquad (2.16)$$

According to Eq. 2.16, the repeated factor $(x-2)^2$ in the denominator of the following rational expression leads to the partial fraction decomposition given by

$$\frac{x+1}{(x-2)^2(x-3)} = \frac{A_1}{x-2} + \frac{A_2}{(x-2)^2} + \frac{A_3}{x-3}$$

(Note that the term $(x-3)$ of the denominator continues to have the single partial fraction indicated by Eq. 2.15.)

Example 2.32

Using the method of partial fractions, determine the following integral.

$$\int \frac{4x}{(x-1)^2(x+1)} \, dx$$

Solution:

The denominator of the integrand is factored into two factors, one of which is repeated. According to Eqs. 2.15 and 2.16, the partial fraction decomposition of the integrand is given by

$$\frac{4x}{(x-1)^2(x+1)} = \frac{A_1}{x-1} + \frac{A_2}{(x-1)^2} + \frac{A_3}{x+1}$$

Clear the fractions.

$$4x = A_1(x-1)(x+1) + A_2(x+1) + A_3(x-1)^2$$
$$= (A_1 + A_3)x^2 + (A_2 - 2A_3)x + A_2 - A_1 + A_3$$

Collect terms with the same power of x to obtain the following system of algebraic equations.

$$0 = A_1 + A_3$$
$$4 = A_2 - 2A_3$$
$$0 = A_2 - A_1 + A_3$$

Use Cramer's rule or elimination to obtain $A_1 = 1$, $A_2 = 2$, and $A_3 = -1$.

Alternatively, substitute, in turn, $x = 1$ and $x = -1$ into the equation.

$$4x = A_1(x-1)(x+1) + A_2(x+1) + A_3(x-1)^2$$

Respectively, this leads to

$$4 = 2A_2$$
$$-4 = 4A_3$$

Solve this simple set of equations to obtain $A_2 = 2$ and $A_3 = -1$. The coefficient A_1 is now found by solving the following single equation obtained previously but this time, using the known values of A_2 and A_3.

$$0 = A_2 - A_1 + A_3$$
$$0 = 2 - A_1 + (-1)$$
$$0 = 1 - A_1$$

It follows that $A_1 = 1$, $A_2 = 2$, and $A_3 = -1$ as before.

The integral now becomes

$$\int\left(\frac{4x}{(x-1)^2(x+1)}\right)dx$$

$$=\int\left[\frac{A_1}{x-1}+\frac{A_2}{(x-1)^2}+\frac{A_3}{x+1}\right]dx$$

$$=\int\left[\frac{1}{x-1}+\frac{2}{(x-1)^2}-\frac{1}{x+1}\right]dx$$

From Table 2.3,

$$\int\left(\frac{4x}{(x-1)^2(x+1)}\right)dx=\ln|x-1|-\frac{2}{x-1}$$
$$-\ln|x+1|+C$$

If, in the expression P/Q, Q contains an *irreducible quadratic factor* (i.e., a quadratic factor that cannot be factored into two real linear factors—more precisely, a factor of the form ax^2+bx+c, where $b^2-4ac<0$), the corresponding partial fraction will take the following form.

$$\frac{Ax+B}{ax^2+bx+c} \qquad (2.17)$$

If that factor is repeated, for example, r times, then instead of the single partial fraction in Eq. 2.17, assign r partial fractions as follows.

$$\frac{P(x)}{(ax^2+bx+c)^r}=\frac{A_1x+B_1}{ax^2+bx+c}+\frac{A_2x+B_2}{(ax^2+bx+c)^2}$$
$$+\cdots+\frac{A_rx+B_r}{(ax^2+bx+c)^r} \qquad (2.18)$$

The constants $A_1,\ldots,A_r$ are found as in Ex. 2.32.

Example 2.33
Determine the following integral.

$$\int\frac{x^2-2}{x(x^2+2)}dx$$

Solution:
The integrand is a proper rational function whose denominator is factored into a quadratic factor, x^2+2, and a linear factor, x. With Eqs. 2.15 and 2.17, the integrand has the following partial fraction decomposition.

$$\frac{x^2-2}{x(x^2+2)}=\frac{A}{x}+\frac{Bx+C}{(x^2+2)}$$

Following the procedure used in Exs. 2.31 to 2.32, find the constants A, B, and C.

$$\frac{x^2-2}{x(x^2+2)}=\frac{A}{x}+\frac{Bx+C}{(x^2+2)}$$
$$x^2-2=A(x^2+2)+(Bx+C)x$$
$$=(A+B)x^2+Cx+2A$$

Compare coefficients of the different powers of x to obtain the following algebraic equations.

$$A+B=1$$
$$C=0$$
$$2A=-2$$

The solution is $A=-1$, $B=2$, and $C=0$.

Alternatively, let $x=0$ in the following equation.

$$x^2-2=A(x^2+2)+(Bx+C)x$$

This leads to the following simpler equation.

$$-2=2A$$

This equation has solution $A=-1$. The values of the constants B and C are obtained using the following equations derived previously.

$$A+B=1$$
$$C=0$$

Once again, $A=-1$, $B=2$, and $C=0$.

The integral now becomes

$$\int\left(\frac{x^2-2}{x(x^2+2)}\right)dx=\int\left[\frac{A}{x}+\frac{Bx+C}{x^2+2}\right]dx$$
$$=\int\left(\frac{-1}{x}+\frac{2x}{x^2+2}\right)dx$$

From Table 2.3,

$$\int\left(\frac{x^2-2}{x(x^2+2)}\right)dx=\int\left(\frac{-1}{x}+\frac{2x}{x^2+2}\right)dx$$
$$=-\ln|x|+\ln|x^2+2|+C$$

The answer can be simplified using the rules for logarithms (see App. 3).

$$\int\frac{x^2-2}{x(x^2+2)}dx=-\ln|x|+\ln|x^2+2|+C$$
$$=\ln\left|\frac{x^2+2}{x}\right|+C$$

(Note that the procedure is similar for definite integrals.)

Example 2.34

Evaluate the following integral.

$$\int_0^{\frac{1}{2}} \frac{dx}{x^4 - 1}$$

Solution:

The integrand is a proper rational function, and the denominator factors into the following product of three factors.

$$x^4 - 1 = (x^2 - 1)(x^2 + 1) = (x - 1)(x + 1)(x^2 + 1)$$

The integrand becomes

$$\frac{1}{x^4 - 1} = \frac{1}{(x - 1)(x + 1)(x^2 + 1)}$$

The partial fractions are obtained from Eqs. 2.15 and 2.17.

$$\frac{1}{(x - 1)(x + 1)(x^2 + 1)} = \frac{A}{x - 1} + \frac{B}{x + 1} + \frac{Ex + F}{x^2 + 1}$$

Clear the fractions by multiplying both sides of this equation by $(x - 1)(x + 1)(x^2 + 1)$.

$$\begin{aligned}
1 &= A(x + 1)(x^2 + 1) + B(x - 1)(x^2 + 1) \\
&\quad + (Ex + F)(x - 1)(x + 1) \\
&= Ax^3 + Ax + Ax^2 + A + Bx^3 + Bx - Bx^2 - B \\
&\quad + Ex^3 - Ex + Fx^2 - F
\end{aligned}$$

Collect terms with the same power of x.

$$\begin{aligned}
1 &= x^3(A + B + E) + x^2(A - B + F) + x(A + B - E) \\
&\quad + (A - B - F)
\end{aligned}$$

Equate coefficients of different powers of x to obtain the following algebraic equations.

$$\begin{aligned}
0 &= A + B + E \\
0 &= A - B + F \\
0 &= A + B - E \\
1 &= A - B - F
\end{aligned}$$

Solve using Cramer's rule or elimination to obtain $A = 1/4$, $B = -1/4$, $E = 0$, and $F = -1/2$.

Alternatively, setting, in turn, $x = -1$ and $x = 1$ in the following equation leads to a simpler set of algebraic equations for the constants A, B, E, and F.

$$\begin{aligned}
1 &= A(x + 1)(x^2 + 1) + B(x - 1)(x^2 + 1) \\
&\quad + (Ex + F)(x - 1)(x + 1)
\end{aligned}$$

Let $x = -1$.

$$1 = -4B$$

Let $x = 1$.

$$1 = 4A$$

It follows that $A = 1/4$ and $B = -1/4$. The values of E and F are found from the following equations obtained above.

$$\begin{aligned}
0 &= A + B + E \\
0 &= A - B + F
\end{aligned}$$

Once again, $A = 1/4$, $B = -1/4$, $E = 0$, and $F = -1/2$.

The integral now becomes

$$\begin{aligned}
\int \frac{dx}{x^4 - 1} &= \int \left(\frac{A}{x - 1} + \frac{B}{x + 1} + \frac{Ex + F}{x^2 + 1} \right) dx \\
&= \int \left(\frac{\frac{1}{4}}{x - 1} + \frac{-\frac{1}{4}}{x + 1} + \frac{-\frac{1}{2}}{x^2 + 1} \right) dx
\end{aligned}$$

Using Table 2.3,

$$\begin{aligned}
\int \frac{dx}{x^4 - 1} &= \frac{1}{4} \ln |x - 1| - \frac{1}{4} \ln |x + 1| - \frac{1}{2} \arctan x + C \\
&= \frac{1}{4} \ln \left| \frac{x - 1}{x + 1} \right| - \frac{1}{2} \arctan x + C
\end{aligned}$$

Finally,

$$\begin{aligned}
\int_0^{\frac{1}{2}} \frac{dx}{x^4 - 1} &= \left[\frac{1}{4} \ln \left| \frac{x - 1}{x + 1} \right| - \frac{1}{2} \arctan x \right]_0^{\frac{1}{2}} \\
&= \left[\frac{1}{4} \ln \left| \frac{-\frac{1}{2}}{\frac{3}{2}} \right| - \frac{1}{2} \arctan \left(\frac{1}{2} \right) \right] \\
&= - \left(\frac{1}{4} \ln \left| \frac{-1}{1} \right| - \frac{1}{2} \arctan (0) \right) \\
&= \left[\frac{1}{4} \ln \frac{1}{3} - \frac{1}{2} \arctan \left(\frac{1}{2} \right) \right] \\
&\quad - \left[\frac{1}{4} \ln 1 - \frac{1}{2} \arctan (0) \right] \\
&= \frac{1}{4} \ln \frac{1}{3} - \frac{1}{2} \arctan \left(\frac{1}{2} \right) \sim -0.506
\end{aligned}$$

PRACTICE PROBLEMS

4. Use integration by parts to determine the following integrals.

(a) $\int \left(\dfrac{\ln x}{x^2} \right) dx$

(b) $\int t^{\frac{1}{2}} \ln t \, dt$

(c) $\int_0^1 x e^{2x} \, dx$

(d) $\int t^2 \sin t \, dt$

5. Use the given substitution to determine the following integrals.

(a) $\int \cos^2 x \sin x \, dx; \ u = \cos x$

(b) $\int x(x^2 + 1)^{25} \, dx; \ u = x^2 + 1$

6. Find a suitable substitution and determine the following integrals.

(a) $\int \left(\dfrac{6x^5 + 10}{\sqrt{x^6 + 10x}} \right) dx$

(b) $\int 4x^3 \cos \left(2x^4 + 10 \right) dx$

(c) $\int_0^4 x \sqrt{x^2 + 9} \, dx$

(d) $\int_{-1}^1 \left(\dfrac{(x+1)}{\sqrt{x^2 + 2x + 1}} \right) dx$

7. Use a trigonometric substitution to determine the following integral.

$$\int \frac{\sqrt{4 - x^2}}{x} \, dx$$

8. Use the method of partial fractions to determine the following integrals.

(a) $\int \dfrac{dx}{x^3 - x}$

(b) $\int \left(\dfrac{x+1}{x^3 - x^2} \right) dx$

(c) $\int \left(\dfrac{x}{(x+1)(x^2 + 1)} \right) dx$

9. Determine $\int \arctan x \, dx$.

FE-Style Exam Problems

In what follows, C is an arbitrary constant.

1. Determine the following indefinite integral.

$$\int \left(\frac{x^2 + x + 1}{\sqrt{x}} \right) dx$$

(A) $\dfrac{2}{5} x^{\frac{5}{2}} + \dfrac{3}{2} x^{\frac{3}{2}} + 2x^{\frac{1}{2}}$

(B) $\dfrac{2}{5} x^{\frac{5}{2}} + \dfrac{2}{3} x^{\frac{3}{2}} + 2x^{\frac{1}{2}} + C$

(C) $\dfrac{2}{5} x^{\frac{5}{2}} + \dfrac{2}{3} x^{\frac{3}{2}} + 2x^{\frac{1}{2}}$

(D) $\dfrac{3}{2} x^{\frac{1}{2}} + \dfrac{1}{2} x^{-\frac{1}{2}} - \dfrac{1}{2} x^{-\frac{3}{2}} + C$

2. Find the area of the region lying under the curve $y = 4x^2 - 4x + 3$, above the x-axis and between the lines $x = 0$ and $x = 2$.

(A) $-\dfrac{25}{3}$

(B) 0

(C) $\dfrac{26}{3}$

(D) $\dfrac{4}{3} x^3 - 2x^2 + 3x$

3. Determine the following indefinite integral.

$$\int \left(\frac{\cos x}{1 + \sin^2 x} \right) dx$$

(A) $\arctan (\sin x) + C$

(B) $x + C$

(C) $\ln |\sin x| + C$

(D) $\arcsin (\tan x) + C$

4. Evaluate the following definite integral.

$$\int_0^{\frac{\pi}{2}} x \sin x \, dx$$

(A) -5

(B) -1

(C) 0

(D) 1

5. Find the area between the curve $y = \sin x$, the x-axis, and the lines $x = 0$ and $x = \pi$.

(A) -2

(B) 2

(C) 3

(D) 15

6. Determine the following indefinite integral.

$$\int (3x - 1)e^x \, dx$$

(A) $e^x(3x - 4) + C$

(B) $e^x(3x - 2) + C$

(C) $\dfrac{x^2}{2}e^x - e^x + C$

(D) 2

7. Determine the following indefinite integral.

$$\int \sin x \ln(\cos x) \, dx \quad [0 < x < \tfrac{\pi}{2}]$$

(A) $\ln(\cos x)(\sin x + C)$

(B) $\cos^2 x + C$

(C) $-\cos x[\ln(\cos x) - 1] + C$

(D) $-\cos x[\ln(\cos x) + 1] + C$

8. Determine the following indefinite integral.

$$\int \left(\frac{4x}{(x^2 - 2x + 1)(x + 1)} \right) dx$$

(A) $\ln(x - 1) - \dfrac{2}{x - 1} - \ln(x + 1) + C$

(B) $\dfrac{4x}{(x^2 - 2x + 1)^2} + C$

(C) $\ln(x - 1) - \dfrac{\frac{1}{2}}{x - 1} - \ln(2x + 1) + C$

(D) $\ln \left| \dfrac{x - 1}{x + 1} \right| - \dfrac{2}{x - 1} + C$

9. Determine the following indefinite integral.

$$\int \left(\frac{\sqrt{2}}{9 + 3x^2} \right) dx$$

(A) $\left(\dfrac{\sqrt{2}}{3\sqrt{3}} \right) \arctan \left(\dfrac{x}{\sqrt{3}} \right) + C$

(B) $\sqrt{2} \ln(9 + 3x^2) + C$

(C) $\left(\dfrac{\sqrt{2}}{3} \right) \arctan \left(\dfrac{x}{\sqrt{3}} \right) + C$

(D) $\left(\dfrac{\sqrt{2}}{\sqrt{3}} \right) \arcsin \left(\dfrac{x}{\sqrt{3}} \right) + C$

10. Evaluate the following definite integral.

$$\int_{-1}^{-\frac{1}{2}} \left(\frac{x + 1}{x^2 + 2x} \right) dx$$

(A) $\dfrac{1}{2} \ln \dfrac{3}{4}$

(B) $2 \ln \dfrac{3}{4}$

(C) 0

(D) e

3
Centroids and Moments of Inertia

One of the most important applications of integral calculus lies in the calculation of certain physical quantities such as the *centroid* or *moment of inertia*, both of which arise frequently in engineering mechanics and mechanics of materials. In this chapter, techniques from Chap. 2 are used to calculate the *centroid of an area* (the analogy of the center of gravity of a homogeneous body) and the *area moment of inertia* for an area bounded by the x-axis, a curve described by the function $y = f(x)$ and the lines $x = a$ and $x = b$.

Figure 3.1 illustrates the region S of the xy-plane.

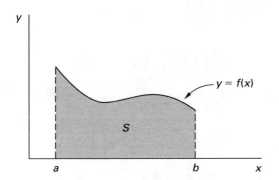

Figure 3.1 The Region S Bounded by the x-Axis, the Curve $y = f(x)$, and the Lines $x = a$ and $x = b$

The *centroid* of the region S is located at position (x_c, y_c) where

$$x_c = \frac{M_y}{A} \qquad (3.1a)$$

$$y_c = \frac{M_x}{A} \qquad (3.1b)$$

In Eq. 3.1, A represents the area of the region S and is defined by

$$A = \int_S dA \qquad (3.2)$$

The quantities M_x and M_y of Eq. 3.1 are called the *first (area) moments of the region S* about the x- and y-axes, respectively. They are defined by

$$M_x = \int_S y\,dA \qquad (3.3a)$$

$$M_y = \int_S x\,dA \qquad (3.3b)$$

The *area integrals* in Eqs. 3.2 and 3.3 are actually *double integrals* and, in general, are very difficult to evaluate. Fortunately, for the specific region S above, M_x and M_y reduce to *single integrals* of the type discussed in Chap. 2. In fact,

$$M_y = \int_a^b x f(x)\,dx$$

$$M_x = \int_a^b \left(\frac{1}{2}\right) [f(x)]^2\,dx$$

$$A = \int_a^b f(x)\,dx$$

From Eqs. 3.1 and 3.2,

$$x_c = \frac{1}{A} \int_a^b x f(x)\,dx \qquad (3.4a)$$

$$y_c = \frac{1}{A} \int_a^b \left(\frac{1}{2}\right) [f(x)]^2\,dx \qquad (3.4b)$$

$$A = \int_a^b f(x)\,dx \qquad (3.4c)$$

The results in Eq. 3.4 can be justified through the theory of double integration as follows. From Eq. 3.3, if the area element dA is described by $dA = dy\,dx$,

$$M_x = \int_S y\,dA = \int_a^b \int_0^{f(x)} y\,dy\,dx$$

$$= \int_a^b \left[\frac{y^2}{2}\right]_0^{f(x)} dx$$

$$= \int_a^b \left(\frac{1}{2}\right) [f(x)]^2\,dx$$

Similarly,

$$M_y = \int_S x\,dA$$

$$= \int_a^b \int_0^{f(x)} x\,dy\,dx$$

$$= \int_a^b [xy]_0^{f(x)}\,dx$$

$$= \int_a^b xf(x)\,dx$$

For the area, A,

$$A = \int_S dA$$

$$= \int_a^b \int_0^{f(x)} dy\,dx$$

$$= \int_a^b [y]_0^{f(x)}\,dx$$

$$= \int_a^b f(x)\,dx$$

Example 3.1

Find the centroid of the area, A, bounded by the curve $y = x^2 + 1$, the x-axis, and the lines $x = \pm 1$.

Solution:

The area, A, is illustrated in the following figure.

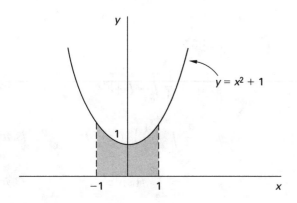

From Eq. 3.4 with $y = f(x) = x^2 + 1$, $a = -1$, and $b = 1$,

$$A = \int_a^b f(x)\,dx = \int_{-1}^1 (x^2 + 1)\,dx$$

$$= \left[\frac{x^3}{3} + x\right]_{-1}^1$$

$$= \left(\frac{1}{3} + 1\right) - \left[\frac{(-1)^3}{3} + (-1)\right] = \frac{8}{3}$$

$$x_c = \frac{1}{A}\int_a^b xf(x)\,dx$$

$$= \left(\frac{1}{\frac{8}{3}}\right)\int_{-1}^1 x(x^2 + 1)\,dx$$

$$= \frac{3}{8}\int_{-1}^1 x(x^2 + 1)\,dx$$

$$= \frac{3}{8}\int_{-1}^1 (x^3 + x)\,dx$$

$$= \left(\frac{3}{8}\right)\left[\frac{x^4}{4} + \frac{x^2}{2}\right]_{-1}^1$$

$$= \left(\frac{3}{8}\right)\left[\left(\frac{1}{4} + \frac{1}{2}\right) - \left(\frac{(-1)^4}{4} + \frac{(-1)^2}{2}\right)\right]$$

$$= \left(\frac{3}{8}\right)\left[\left(\frac{1}{4} + \frac{1}{2}\right) - \left(\frac{1}{4} + \frac{1}{2}\right)\right] = 0$$

Finally,

$$y_c = \frac{1}{A}\int_a^b \left(\frac{1}{2}\right)[f(x)]^2\,dx$$

$$= \left(\frac{1}{\frac{8}{3}}\right)\left(\frac{1}{2}\right)\int_{-1}^1 (x^2 + 1)^2\,dx$$

$$= \left(\frac{3}{8}\right)\left(\frac{1}{2}\right)\int_{-1}^1 (x^2 + 1)^2\,dx$$

$$= \frac{3}{16}\int_{-1}^1 (x^4 + 2x^2 + 1)\,dx$$

$$= \left(\frac{3}{16}\right)\left[\frac{x^5}{5} + \frac{2}{3}x^3 + x\right]_{-1}^1$$

$$= \left(\frac{3}{16}\right)\left[\left(\frac{1}{5} + \frac{2}{3} + 1\right)\right.$$

$$\left. - \left(\frac{(-1)^5}{5} + \left(\frac{2}{3}\right)(-1)^3 + (-1)\right)\right]$$

$$= \left(\frac{3}{16}\right)\left(\frac{28}{15} + \frac{28}{15}\right) = \frac{7}{10}$$

It follows that $(x_c, y_c) = (0, 7/10)$.

Example 3.2

Find the centroid of the area, A, bounded by the curve $y = \cos x$ and the coordinate axes in the first quadrant.

Solution:

The area, A, is illustrated in the following figure where the appropriate x-limits of integration are $a = 0$ and $b = \pi/2$.

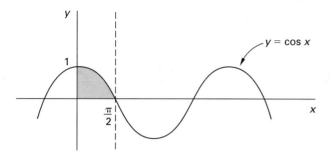

From Eq. 3.4, with $f(x) = \cos x$, $a = 0$, and $b = \pi/2$,

$$A = \int_0^{\frac{\pi}{2}} \cos x \, dx$$

$$= \left[\sin x \right]_0^{\frac{\pi}{2}}$$

$$= \sin\left(\frac{\pi}{2}\right) - \sin(0)$$

$$= 1 - 0 = 1$$

$$x_c = \frac{1}{A} \int_a^b x f(x) dx$$

$$= \frac{1}{1} \int_0^{\frac{\pi}{2}} x \cos x \, dx$$

$$= \int_0^{\frac{\pi}{2}} x \cos x \, dx$$

This definite integral is evaluated using integration by parts. From Eq. 2.9 with $f(x) = x$, $h(x) = \cos x$, and $H(x) = \sin x$,

$$x_c = \int_0^{\frac{\pi}{2}} x \cos x \, dx$$

$$= \left[x \sin x \right]_0^{\frac{\pi}{2}} - \int_0^{\frac{\pi}{2}} (\sin x)(1) \, dx$$

$$= \left[\frac{\pi}{2} \sin\left(\frac{\pi}{2}\right) - (0)\sin(0) \right] - \left[-\cos x \right]_0^{\frac{\pi}{2}}$$

$$= \left[\left(\frac{\pi}{2}\right)(1) - 0 \right] + \left[\cos\left(\frac{\pi}{2}\right) - \cos 0 \right]$$

$$= \frac{\pi}{2} - 1$$

Also from Eq. 3.4, with $f(x) = \cos x$, $a = 0$, and $b = \pi/2$,

$$y_c = \frac{1}{A} \int_a^b \left(\frac{1}{2}\right) [f(x)]^2 dx$$

$$= \left(\frac{1}{1}\right)\left(\frac{1}{2}\right) \int_0^{\frac{\pi}{2}} \cos^2 x \, dx$$

$$= \frac{1}{2} \int_0^{\frac{\pi}{2}} \cos^2 x \, dx$$

From Table 2.2,

$$y_c = \frac{1}{2} \int_0^{\frac{\pi}{2}} \cos^2 x \, dx$$

$$= \left(\frac{1}{2}\right)\left[\frac{x}{2} + \frac{1}{4}\sin 2x \right]_0^{\frac{\pi}{2}}$$

$$= \left(\frac{1}{2}\right)\left[\left(\frac{\pi}{4} + \frac{1}{4}\sin \pi\right) - \left(0 + \frac{1}{4}\sin 0\right) \right]$$

$$= \left(\frac{1}{2}\right)\left[\left(\frac{\pi}{4} + 0\right) - 0 \right] = \frac{\pi}{8}$$

It follows that $(x_c, y_c) = \pi/2 - (1, \pi/8)$.

··

Centroids for most of the common shapes arising in applications are usually obtained from tables. Only in very special cases is it necessary to resort to Eq. 3.4 (or indeed Eqs. 3.1 through 3.3) directly to locate (x_c, y_c).

Suppose that the region S is bounded by the y-axis, a curve represented by the function $x = g(y)$, and the lines $y = c$ and $y = d$, as illustrated in Fig. 3.2.

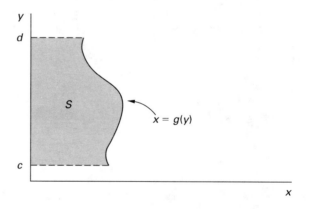

Figure 3.2 The Region S Bounded by the y–Axis,
the Curve $x = g(y)$, and the Lines $y = c$ and $y = d$

Using the theory of double integration (noting that the area element is now $dA = dx \, dy$, the x-limits of integration are $x = 0$ to $x = g(y)$, and the y-limits of integration are $y = c$ to $y = d$), it can be shown that instead of the formulas in Eq. 3.4, Eqs. 3.1 through 3.3 lead to

$$x_c = \frac{1}{A} \int_c^d \left(\frac{1}{2}\right) [g(y)]^2 dy \qquad (3.5a)$$

$$y_c = \frac{1}{A} \int_c^d y g(y) dy \qquad (3.5b)$$

$$A = \int_c^d g(y) dy \qquad (3.5c)$$

Example 3.3

Consider the area bounded by the lines $y = f(x) = (2)(3 - x)$, $x = 0$, and $y = 0$ as illustrated in the following figure.

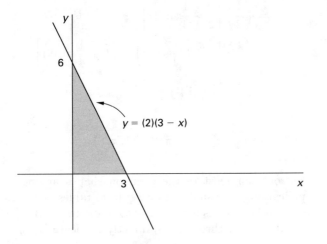

Solution:

This question can be solved in two ways: using either Eq. 3.4 or Eq. 3.5. First, use Eq. 3.4. From the following figure, the area, A, can be regarded as the area bounded by the x-axis, the curve $y = f(x) = (2)(3 - x)$, and the lines $x = 0$ and $x = 3$.

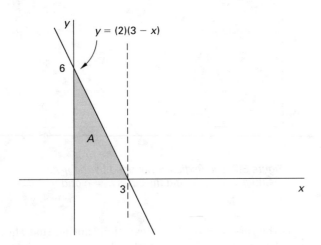

Using Eq. 3.4 with $f(x) = (2)(3 - x)$, $a = 0$, and $b = 3$,

$$A = \int_0^3 (2)(3 - x)dx$$

$$= \left[6x - x^2 \right]_0^3 = 9$$

$$x_c = \left(\frac{1}{9} \right) \int_0^3 (x)(2)(3 - x)dx$$

$$= \left(\frac{1}{9} \right) \left[\frac{6x^2}{2} - \frac{2x^3}{3} \right]_0^3$$

$$= \left(\frac{1}{9} \right)(9) = 1$$

$$y_c = \left(\frac{1}{9} \right) \left(\frac{1}{2} \right) \int_0^3 (4)(3 - x)^2 dx$$

$$= \left(\frac{1}{18} \right)(4) \left[9x - \frac{6x^2}{2} + \frac{x^3}{3} \right]_0^3$$

$$= \left(\frac{2}{9} \right)(9) = 2$$

Alternatively, regard the area, A, as the area enclosed by the y-axis, the curve $x = g(y) = (1/2)(6 - y)$ (obtained from the equation $y = f(x) = (2)(3 - x)$ rewritten as a function of y), and the lines $y = 0$ and $y = 6$, as illustrated in the following figure.

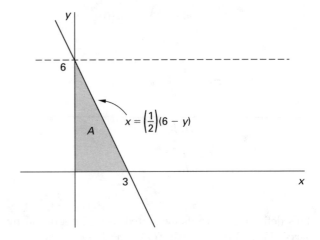

Then, using Eq. 3.5 with $x = g(y) = (1/2)(6 - y)$, $c = 0$, and $d = 6$,

$$A = \int_0^6 \left(\frac{1}{2} \right)(6 - y)dy$$

$$= \left(\frac{1}{2} \right) \left[6y - \frac{y^2}{2} \right]_0^6$$

$$= \left(\frac{1}{2} \right)(18) = 9$$

$$x_c = \frac{1}{A} \int_0^6 \left(\frac{1}{2} \right) [g(y)]^2 dy$$

$$= \left(\frac{1}{9} \right) \int_0^6 \left(\frac{1}{2} \right) \left(\frac{1}{4} \right)(6 - y)^2 dy$$

This integral can be evaluated using Table 2.3.

$$x_c = \left(\frac{1}{9} \right) \left(\frac{1}{8} \right) \left[-\frac{(6 - y)^3}{3} \right]_0^6$$

$$= \left(\frac{1}{72} \right) \left(-\frac{1}{3} \right)(0 - 216) = 1$$

Finally,

$$y_c = \frac{1}{A} \int_0^6 y \left[\left(\frac{1}{2} \right) (6-y) \right] dy$$

$$= \left(\frac{1}{9} \right) \left(\frac{1}{2} \right) \int_0^6 y(6-y) dy$$

$$= \left(\frac{1}{18} \right) \left[\frac{6y^2}{2} - \frac{y^3}{3} \right]_0^6$$

$$= \left(\frac{1}{18} \right) (108 - 72) = 2$$

In each case, $(x_c, y_c) = (1, 2)$ and $A = 9$.

Moments of Inertia

Second area moments of the region S, such as $\int_S x^2 dA$, are referred to as *(area) moments of inertia for the region S (or area A).* The terminology "moment of inertia," as used here, is actually a misnomer: It is used because of the similarity with integrals of the same form (mass moments of inertia) related to mass. Nevertheless, integrals of the type $\int_S x^2 dA$ merit individual attention because they arise frequently in applications (e.g., when calculating a mass moment of inertia for a body with constant density).

The *area moments of inertia*, I_x and I_y, of the region S (or area A) with respect to the x- and y-axes are defined by

$$I_x = \int_S y^2 dA \qquad (3.6a)$$

$$I_y = \int_S x^2 dA \qquad (3.6b)$$

The integrals in Eq. 3.6 are again double integrals and, as such, are difficult to evaluate for a general region S of the xy-plane. However, for the specific region S in Fig. 3.1, the formulas for I_x and I_y again reduce to simpler formulas involving single integrals of the type discussed in Chap. 2.

When the region S is bounded by the x-axis, the curve $y = f(x)$, and the lines $x = a$ and $x = b$, as in Fig. 3.1,

$$I_x = \int_a^b \left(\frac{1}{3} \right) (f(x))^3 dx \qquad (3.7a)$$

$$I_y = \int_a^b x^2 f(x) dx \qquad (3.7b)$$

Similarly, when the region S is bounded by the y-axis, the curve $x = g(y)$, and the lines $y = c$ and $y = d$, as in Fig. 3.2,

$$I_x = \int_c^d y^2 g(y) dy \qquad (3.8a)$$

$$I_y = \int_c^d \left(\frac{1}{3} \right) (g(y))^3 dy \qquad (3.8b)$$

Example 3.4

Find I_x and I_y for the region S bounded by the y-axis, the line $y = 8$, and the curve $y^2 = 8x$.

Solution:

The region S is illustrated as follows.

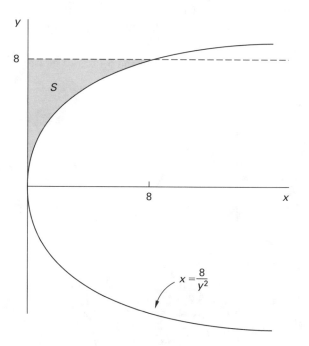

Use Eq. 3.8 with $x = g(y) = y^2/8$ to find I_x and I_y. From the previous illustration the y-limits of integration are $c = 0$ and $d = 8$. From Eq. 3.8,

$$I_x = \int_0^8 y^2 g(y) dy$$

$$= \int_0^8 (y^2) \left(\frac{y^2}{8} \right) dy = \frac{1}{8} \int_0^8 y^4 dy$$

$$= \left(\frac{1}{8} \right) \left[\frac{y^5}{5} \right]_0^8$$

$$= \left(\frac{1}{8} \right) \left(\frac{(8)^5}{5} - 0 \right) = \frac{4096}{5}$$

Similarly,

$$I_y = \int_0^8 \left(\frac{1}{3}\right) \left(g(y)\right)^3 dy$$

$$= \frac{1}{3} \int_0^8 \left(\frac{y^2}{8}\right)^3 dy$$

$$= \frac{1}{(3)(8)^3} \int_0^8 y^6 dy$$

$$= \left[\frac{1}{(3)(8)^3}\right] \left[\frac{y^7}{7}\right]_0^8$$

$$= \left[\frac{1}{(3)(8)^3}\right] \left[\frac{(8)^7}{7} - 0\right]$$

$$= \left(\frac{1}{21}\right) (8)^4 = \frac{4096}{21}$$

Example 3.5

Find I_x and I_y for the region S bounded by the x-axis, the curve $y = \sin x$, and the lines $x = 0$ to $x = \pi$.

Solution:

The region S is illustrated in the following figure.

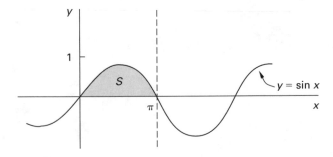

Use Eq. 3.7 with $f(x) = \sin x$ to find I_x and I_y. From the figure, the x-limits of integration are $a = 0$ and $b = \pi$. From Eq. 3.7,

$$I_x = \int_a^b \left(\frac{1}{3}\right) [f(x)]^3 dx = \frac{1}{3} \int_0^\pi \sin^3 x \, dx$$

To evaluate this integral, first use a trigonometric identity to put the integrand into the form $\int f(g(x))g'(x)dx$; then use the substitution rule.

$$\sin^3 x = (\sin x)(\sin^2 x)$$

$$= (\sin x)(1 - \cos^2 x)$$

The integral becomes

$$I_x = \frac{1}{3} \int_0^\pi (\sin x)(1 - \cos^2 x)dx$$

$$= \frac{1}{3} \int_0^\pi (1 - \cos^2 x)(\sin x \, dx)$$

Next use the substitution rule for integrals of the form $\int f(g(x))g'(x)dx$. Make the substitution $u = \cos x$ since the differential $du = -\sin x \, dx$ (apart from the factor -1) appears in the integral. The integrand becomes

$$(1 - \cos^2 x)(\sin x \, dx) = (1 - u^2)(-du)$$

Since the integral is a definite integral, the limits must also be changed according to the substitution for u: When $x = 0$, $u = \cos 0 = 1$; when $x = \pi$, $u = \cos \pi = -1$.

Consequently, the integral becomes

$$I_x = \frac{1}{3} \int_1^{-1} (1 - u^2)(-du)$$

$$= -\frac{1}{3} \int_1^{-1} (1 - u^2) \, du$$

The limits in this integral are "upside down": The lower limit should be less than the upper limit. The following property of definite integrals is used to correct this situation. For an integrable function f,

$$\int_a^b f(u)du = -\int_b^a f(u)du$$

Consequently,

$$I_x = \left(-\frac{1}{3}\right) \left[-\int_{-1}^1 (1 - u^2)du\right]$$

$$= \frac{1}{3} \int_{-1}^1 (1 - u^2)du$$

Finally, using the fundamental theorem of calculus,

$$I_x = \frac{1}{3} \int_{-1}^1 (1 - u^2) \, du$$

$$= \left(\frac{1}{3}\right) \left[u - \frac{u^3}{3}\right]_{-1}^1$$

$$= \left(\frac{1}{3}\right) \left[\left(1 - \frac{(1)^3}{3}\right) - \left((-1) - \frac{(-1)^3}{3}\right)\right]$$

$$= \left(\frac{1}{3}\right) \left[\left(\frac{2}{3}\right) - \left(-\frac{2}{3}\right)\right] = \frac{4}{9}$$

Similarly, from Eq. 3.7,

$$I_y = \int_a^b x^2 f(x)dx = \int_0^\pi x^2 \sin x \, dx$$

To evaluate this integral, apply integration by parts. Use Eq. 2.9 with $f(x) = x^2$, $h(x) = \sin x$, and $H(x) = -\cos x$.

$$I_y = \int_0^\pi x^2 \sin x \, dx$$

$$= \left[x^2(-\cos x) \right]_0^\pi - \int_0^\pi (-\cos x)(2x) \, dx$$

$$= -(\pi^2 \cos \pi - 0) + 2 \int_0^\pi (\cos x)(x) \, dx$$

$$= \pi^2 + 2 \int_0^\pi (\cos x)(x) \, dx$$

Apply integration by parts to this integral. Use Eq. 2.9 with $f(x) = x$, $h(x) = \cos x$, and $H(x) = \sin x$.

$$I_y = \pi^2 + 2 \int_0^\pi (\cos x)(x) \, dx$$

$$= \pi^2 + (2) \left(\left[x \sin x \right]_0^\pi - \int_0^\pi (\sin x)(1) \, dx \right)$$

$$= \pi^2 + (2) \left[(\pi \sin \pi - 0) - \left[-\cos x \right]_0^\pi \right]$$

$$= \pi^2 + (2)[0 + (\cos \pi - \cos 0)]$$

$$= \pi^2 + (2)(-1 - 1) = \pi^2 - 4$$

The moment of inertia taken with respect to an axis passing through the centroid (*centroidal axis*) of the area A is known as the *centroidal moment of inertia* and is denoted by I_c. If I_c is known, the moment of inertia with respect to any axis parallel to the centroidal axis is easily found using the *parallel axis*, or *transfer axis*, *theorem*, which follows:

$$I_{\text{parallel axis}} = I_c + Ad^2$$

d is the perpendicular distance between the centroidal axis and the parallel axis in question.

Moments of inertia for most of the basic shapes arising in applications can be found from tables. Only in very special circumstances is it necessary to resort to direct calculation using Eqs. 3.6 through 3.8.

PRACTICE PROBLEMS

1. Find the centroid and the moments of inertia, I_x and I_y, for each of the following regions, S.

   (a) S is the region in the first quadrant bounded by the parabola $y = x^2 + 1$ and the line $x = \sqrt{2}$.

   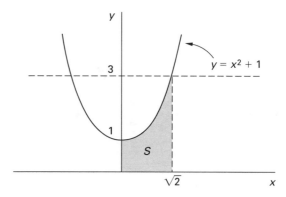

   (b) S is the region in the first quadrant bounded by the line $y = x$, the y-axis, and the line $y = 2$.

   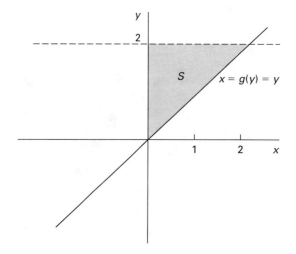

4

Differential Equations

A *differential equation* involves an unknown function and one or more of its derivatives. There are basically two types of differential equations: ordinary differential equations and partial differential equations. *Ordinary differential equations* involve ordinary derivatives since the unknown function depends on a single variable. *Partial differential equations* involve partial derivatives since the unknown function depends on several variables. This chapter is concerned with ordinary differential equations.

The following equation is an example of an ordinary differential equation for the unknown function $y(x)$ of the single variable x.

$$(3)\left(\frac{d^2y}{dx^2}\right) + (2)\left(\frac{dy}{dx}\right) + y = \sin x$$

The variable y is referred to as the *dependent variable*, while x is referred to as the *independent variable*.

The *order* of a differential equation is the order of the highest derivative appearing in the differential equation. For example, the differential equation just given is of order 2 (second order).

A *solution* of the differential equation is any function $y(x)$ satisfying the differential equation.

When solving a differential equation, the objective is to retrieve the solution $y(x)$ from the differential equation. The expression for $y(x)$ need not be explicit, but it cannot contain derivatives. (If it does, the original differential equation has not been solved, but merely replaced by another differential equation).

The retrieval of a function from an equation involving its derivatives will, inevitably, involve integration. For this reason, it is expected that any solution of a differential equation will involve arbitrary constants of integration. This means that a single differential equation will have many solutions (in the same way that a function has many antiderivatives). The collection of all the solutions of a differential equation into a general form is known as the *general solution* of the differential equation (the analogue of the most general antiderivative of a function introduced in Chap. 2).

Often additional data known as boundary conditions or initial conditions are specified. The term *boundary condition* is used when the independent variable, x, represents some physical quantity such as distance. For example, if y depends on x, a boundary condition might take the form $y = 1$ when $x = 0$ or $y(0) = 1$.

The term *initial condition* is usually reserved for problems where the independent variable is time, t. For example, if y depends on t, an initial condition might take the form $y = 3$ when $t = 0$ or $y(0) = 3$, meaning "initially $y = 3$."

Boundary or initial conditions are used to evaluate the constants of integration appearing in the general solution. This leads to a particular, or specific, solution of the differential equation. Any problem that includes a differential equation and boundary or initial conditions is known, respectively, as a *boundary value* or *initial value* problem.

Consider the following expression.

$$y(x) = c_1 \sin x + c_2 \cos x$$

Here, c_1 and c_2 are arbitrary constants. On differentiating y twice with respect to x,

$$\frac{dy}{dx} = c_1 \cos x - c_2 \sin x$$
$$\frac{d^2y(x)}{dx^2} = -c_1 \sin x - c_2 \cos x$$

Consequently,

$$\frac{d^2y(x)}{dx^2} + y(x) = 0$$

In other words, the function $y(x) = c_1 \sin x + c_2 \cos x$ is a solution of the ordinary differential equation

$(d^2y(x))/dx^2 + y(x) = 0$ for any constants c_1 and c_2. In fact, $y(x) = c_1 \sin x + c_2 \cos x$ is the general solution of the differential equation. (The fact that this is the general solution, along with methods for finding the general solution starting from the differential equation, will be discussed in the next section.)

Suppose the following additional (boundary) data are specified.

$$y = 0 \text{ when } x = 0 \quad [\text{i.e., } y(0) = 0]$$

$$\frac{dy}{dx} = 1 \text{ when } x = 0 \quad [\text{i.e., } \left(\frac{dy}{dx}\right)(0) = 1]$$

In such cases, the following equations must be satisfied.

$$y(0) = c_1 \sin 0 + c_2 \cos 0 = 0$$

$$\left(\frac{dy}{dx}\right)(0) = c_1 \cos 0 - c_2 \sin 0 = 1$$

These equations have the solution $c_1 = 1$ and $c_2 = 0$. In this particular case the general solution reduces to

$$y(x) = (1)(\sin x) + (0)(\cos x)$$
$$= \sin x$$

Only in very special cases is it possible to solve an ordinary differential equation exactly and explicitly. One such case involves the class of *linear ordinary differential equations* with constant coefficients.

A *linear N^{th}-order ordinary differential equation* has the following general form.

$$b_N \left(\frac{d^N y(x)}{dx^N}\right) + \cdots + b_1 \left(\frac{dy(x)}{dx}\right) + b_0 y(x) = f(x) \quad \textbf{(4.1)}$$

In Eq. 4.1, N is a positive integer, $b_0, b_1, \ldots, b_N$ are coefficients that may depend on x (but not on y), and $f(x)$ is a known (forcing) function of x that, in general, is non-zero. Any ordinary differential equation deviating from the form given in Eq. 4.1 is referred to as *nonlinear*.

Following are examples of ordinary differential equations that are all linear since they fit the form of Eq. 4.1.

$$(12)\left(\frac{d^3 y}{dx^3}\right) + (3)\left(\frac{d^2 y}{dx^2}\right) + y = x^2$$

$$(3x)\left(\frac{d^2 y}{dx^2}\right) + (2)(\sin x)\left(\frac{dy}{dx}\right) + y = \tan x$$

$$(2)\left(\frac{dy}{dx}\right) + y = \ln x$$

Now consider the following ordinary differential equation.

$$(3)\left(\frac{d^2 y}{dx^2}\right) + \sin y \left(\frac{dy}{dx}\right)^2 + y = e^x$$

This equation is nonlinear since it deviates from Eq. 4.1 in that dy/dx is raised to the power 2 and $\sin y$ appears as a coefficient.

The linear differential equation, Eq. 4.1, is amenable to an exact solution for the following reason: Its complete or general solution $y(x)$ decomposes into two simpler parts:

$$y(x) = y_h(x) + y_p(x) \quad \textbf{(4.2)}$$

In Eq. 4.2, the term $y_h(x)$ is the general solution of the *homogeneous equation* associated with Eq. 4.1. That is, $y_h(x)$ is the general solution of Eq. 4.1 with $f(x) \equiv 0$,

$$b_N \left(\frac{d^N y_h(x)}{dx^N}\right) + \cdots + b_1 \left(\frac{dy_h(x)}{dx}\right) + b_0 y_h(x) = 0 \quad \textbf{(4.3)}$$

The term $y_h(x)$ is referred to as the *complementary solution* since it complements $y_p(x)$.

The term $y_p(x)$ is any specific solution of the *nonhomogeneous equation*, Eq. 4.1 (with $f(x) \neq 0$), which is either already known or readily available. The term $y_p(x)$ is known as the *particular solution*.

From Eq. 4.2 it follows that in order to solve Eq. 4.1, two things are required: first, a method for finding $y_h(x)$—that is, a method for solving the associated homogeneous equation, Eq. 4.3; and second, a good supply of particular solutions $y_p(x)$ or a way of obtaining such solutions.

In fact, there is a simple method for obtaining $y_h(x)$ and $y_p(x)$ in the specific case when the coefficients $b_0, b_1, \ldots, b_N$ in Eqs. 4.1 and 4.3 are *constant*. With this in mind, in what follows it is assumed that the coefficients $b_0, b_1, \ldots, b_N$ are constant. That is, the remainder of this chapter will be concerned exclusively with linear ordinary differential equations with constant coefficients.

In the next section, methods for finding $y_h(x)$ are discussed. Subsequently, you will be shown how to obtain $y_p(x)$ for a given non-zero forcing function $f(x)$. The decomposition in Eq. 4.2 is then used to obtain the general solution of Eq. 4.1.

1 Linear Homogeneous Differential Equations with Constant Coefficients

Consider the Nth-order linear homogeneous differential equation with constant coefficients

$$b_N \left(\frac{d^N y(x)}{dx^N}\right) + \cdots + b_1 \left(\frac{dy(x)}{dx}\right) + b_0 y(x) = 0 \quad \textbf{(4.4)}$$

$$[b_0, b_1, \ldots, b_N \text{ are constant.}]$$

Associated with this differential equation is a polynomial equation obtained by replacing each of derivatives $[d^N y(x)]/(dx^N)$, where $N = 1, 2, \ldots$, in Eq. 4.4, by the powers r^N of a new variable r. The term $b_0 y(x)$ in Eq. 4.4 is replaced by $b_0 r^0 = b_0$. That is, the associated polynomial equation is given by

$$b_N r^N + b_{N-1} r^{N-1} + \cdots + b_1 r + b_0 = 0 \qquad \textit{(4.5)}$$

Equation 4.5 is known as the *characteristic equation* associated with Eq. 4.4. The solutions of Eq. 4.5 are intimately connected with the general solution of the differential equation, Eq. 4.4.

Since Eq. 4.5 is a polynomial equation, it has N solutions, some of which may be real, some complex, and some repeated. These solutions are found by obtaining the roots of the *characteristic polynomial*:

$$P(r) = b_N r^N + b_{N-1} r^{N-1} + \cdots + b_1 r + b_0 \qquad \textit{(4.6)}$$

In terms of the characteristic polynomial, the characteristic equation (Eq. 4.5) can be written as

$$P(r) = b_N r^N + b_{N-1} r^{N-1} + \cdots + b_1 r + b_0 = 0 \quad \textit{(4.7)}$$

Example 4.1

What is the characteristic equation associated with the following homogeneous differential equation?

$$\frac{d^3 y}{dx^3} - (6)\left(\frac{d^2 y}{dx^2}\right) + (11)\left(\frac{dy}{dx}\right) - 6y = 0$$

Solution:

As in Eq. 4.7, the characteristic equation is obtained by replacing each of the derivatives by r^N where N is the order of the derivative.

$$P(r) = r^3 - 6r^2 + 11r - 6 = 0$$

The polynomial $P(r)$ can be factored into a product of three factors.

$$P(r) = (r - 1)(r - 2)(r - 3)$$

The characteristic equation now takes the form

$$(r - 1)(r - 2)(r - 3) = 0$$

This equation is satisfied by $r = 1, 2, 3$.

Example 4.2

Solve the characteristic equation associated with the following differential equation.

$$\frac{d^2 y}{dx^2} + (2)\left(\frac{dy}{dx}\right) + y = 0$$

Solution:

As in Eq. 4.7, the associated characteristic equation is given by

$$\begin{aligned} P(r) &= r^2 + 2r + 1 \\ &= (r + 1)^2 \\ &= 0 \end{aligned}$$

This equation has solutions $r = -1, -1$.

Example 4.3

Solve the characteristic equation associated with the following differential equation.

$$\frac{d^2 y}{dx^2} + (2)\left(\frac{dy}{dx}\right) + 2y = 0$$

Solution:

As in Eq. 4.7, the associated characteristic equation is

$$P(r) = r^2 + 2r + 2 = 0$$

According to the formula for the roots of a quadratic equation (see App. 3), $P(r) = r^2 + 2r + 2$ has the following roots.

$$\begin{aligned} r &= \frac{-2 \pm \sqrt{4 - (4)(2)}}{2} \\ &= \frac{-2 \pm \sqrt{-4}}{2} \\ &= \frac{-2 \pm \sqrt{4i^2}}{2} \end{aligned}$$

Here, $i^2 = -1$. Simplifying,

$$r = -1 \pm i$$

The solutions of the characteristic equation are therefore

$$r = -1 \pm i$$

The form of the general solution of the differential equation, Eq. 4.4, is dependent on the nature of the solutions of the characteristic equation.

For *real* solutions of the characteristic equation (Eq. 4.5), there are two cases to consider.

Case 1:

The characteristic equation (Eq. 4.5) has N real and distinct solutions, $r_1, r_2, \ldots, r_N$. In this case, Eq. 4.4 has the following general solution.

$$y(x) = c_1 e^{r_1 x} + c_2 e^{r_2 x} + \cdots + c_N e^{r_N x} \qquad \textit{(4.8)}$$

$$[c_1, c_2, \ldots, c_N \text{ are arbitrary constants.}]$$

Example 4.4

Determine the general solution of the following differential equation.

$$\frac{d^2y}{dx^2} - (2)\left(\frac{dy}{dx}\right) - 3y = 0$$

Solution:

First write down the associated characteristic equation. As in Eq. 4.5, the characteristic equation is given by

$$r^2 - 2r - 3 = 0$$

Factor the left-hand side of this equation.

$$(r - 3)(r + 1) = 0$$

This equation is satisfied by $r = -1, 3$. It follows that the solutions of the characteristic equation are real and distinct. From Eq. 4.8 with $r_1 = -1$ and $r_2 = 3$, the general solution of the differential equation is given by

$$y(x) = c_1 e^{-x} + c_2 e^{3x} \quad [c_1 \text{ and } c_2 \text{ are arbitrary constants.}]$$

Example 4.5

Determine the general solution of the following differential equation.

$$\frac{d^2y}{dx^2} + \frac{dy}{dx} = 0$$

Solution:

First write down the associated characteristic equation. From Eq. 4.5, the characteristic equation is given by

$$r^2 + r = 0$$

Factor the left-hand side of this equation.

$$r(r + 1) = 0$$

This equation is satisfied by $r = -1, 0$. It follows that the solutions of the characteristic equation are real and distinct. From Eq. 4.8 with $r_1 = -1$ and $r_2 = 0$, the general solution of the differential equation is given by

$$y(x) = c_1 e^{-x} + c_2 e^{0x}$$
$$= c_1 e^{-x} + c_2 \quad [c_1 \text{ and } c_2 \text{ are arbitrary constants.}]$$

Example 4.6

Solve the differential equation of Ex. 4.1.

$$\frac{d^3y}{dx^3} - (6)\left(\frac{d^2y}{dx^2}\right) + (11)\left(\frac{dy}{dx}\right) - 6y = 0$$

Solution:

The associated characteristic equation is given by

$$r^3 - 6r^2 + 11r - 6 = 0$$

By Ex. 4.1, this has three real and distinct solutions: $r = 1, 2, 3$. From Eq. 4.8, with $r_1 = 1$, $r_2 = 2$, and $r_3 = 3$, the general solution of the ordinary differential equation is given by

$$y(x) = c_1 e^x + c_2 e^{2x} + c_3 e^{3x} \quad \begin{bmatrix} c_1, c_2, \text{ and } c_3 \text{ are} \\ \text{arbitrary constants.} \end{bmatrix}$$

The second case is concerned with repeated real solutions of the characteristic equation, Eq. 4.5.

Case 2:

The characteristic equation (Eq. 4.5) has a real solution r, repeated N times (a solution r of multiplicity N). In this case, Eq. 4.4 has the following general solution.

$$y(x) = \left(c_1 + c_2 x + \cdots + c_N x^{N-1}\right) e^{rx} \qquad (4.9)$$

$$[c_1, c_2, \ldots, c_N \text{ are arbitrary constants.}]$$

Example 4.7

Determine the general solution of the following differential equation.

$$\frac{d^2y}{dx^2} + (2)\left(\frac{dy}{dx}\right) + y = 0$$

Solution:

First write down the associated characteristic equation. As in Eq. 4.5, the characteristic equation is given by

$$r^2 + 2r + 1 = 0$$

Factor the left-hand side of this equation.

$$(r + 1)^2 = 0$$

This equation is satisfied by $r = -1, -1$. It follows that the characteristic equation has a real, repeated solution of $r = -1, -1$. From Eq. 4.9 with $r = -1$ and $N = 2$, the general solution of the differential equation is

$$y(x) = (c_1 + c_2 x)e^{-x} \quad [c_1 \text{ and } c_2 \text{ are arbitrary constants.}]$$

Example 4.8

Determine the general solution of the following differential equation.

$$\frac{d^2y}{dx^2} = 0$$

Solution:

First write down the associated characteristic equation. As in Eq. 4.5, the characteristic equation is given by

$$r^2 = 0$$

This equation is satisfied by $r = 0, 0$. It follows that the characteristic equation has a real, repeated solution of $r = 0, 0$. From Eq. 4.9 with $r = 0$ and $N = 2$, the general solution of the differential equation is

$$y(x) = (c_1 + c_2 x)e^{0x} = c_1 + c_2 x \qquad \begin{bmatrix} c_1 \text{ and } c_2 \text{ are} \\ \text{arbitrary constants.} \end{bmatrix}$$

Note that this general solution could also have been obtained by direct integration of the differential equation.

$$\frac{d^2 y(x)}{dx^2} = 0$$

$$\frac{dy(x)}{dx} = c_2$$

$$y(x) = c_1 + c_2 x$$

Example 4.9

Determine the general solution of the following differential equation.

$$\frac{d^3 y}{dx^3} - (2)\left(\frac{d^2 y}{dx^2}\right) + (4)\left(\frac{dy}{dx}\right) - 8y = 0$$

Solution:

First write down the associated characteristic equation. As in Eq. 4.5, the characteristic equation is given by

$$r^3 - 2r^2 + 4r - 8 = 0$$

Factor the left-hand side of this equation.

$$(r - 2)^3 = 0$$

This equation is satisfied by $r = 2, 2, 2$. It follows that the characteristic equation has a real, repeated solution of $r = 2, 2, 2$. From Eq. 4.9 with $r = 2$ and $N = 3$, the general solution of the differential equation is

$$y(x) = (c_1 + c_2 x + c_3 x^2)e^{2x} \qquad \begin{bmatrix} c_1, c_2, \text{ and } c_3 \text{ are} \\ \text{arbitrary constants.} \end{bmatrix}$$

It remains to consider the case when the characteristic equation has complex (nonreal) solutions. To this end, note that since the coefficients $b_0, b_1, \ldots, b_N$ of the differential equation in Eq. 4.4 and its characteristic equation, Eq. 4.5, are assumed to be real, any complex (nonreal) solutions of the characteristic equation will occur in conjugate pairs. This means that any complex (nonreal) solutions of the characteristic equation will take the form $\alpha \pm i\beta$ where α and β are real and $i = \sqrt{-1}$. In other words, the characteristic equation cannot have an odd number of complex solutions—only an even number, all occurring in conjugate pairs, $\alpha \pm i\beta$.

When the characteristic equation does, in fact, have complex solutions, the form of the solution of the differential equation in Eq. 4.4 differs significantly from that in Eqs. 4.8 and 4.9.

Case 3:

Suppose that the characteristic equation has an unrepeated pair of complex conjugate solutions of the form $\alpha \pm i\beta$ with $\beta \neq 0$. Then the corresponding part of a general solution of Eq. 4.4 is of the form

$$e^{\alpha x}(c_1 \cos \beta x + c_2 \sin \beta x) \qquad (4.10)$$

$$[c_1 \text{ and } c_2 \text{ are arbitrary constants.}]$$

The result in Case 3 has significant consequences for the general solution of a *second-order* homogeneous linear differential equation (Eq. 4.4 with $N = 2$) whose associated characteristic equation (Eq. 4.5 with $N = 2$) has complex solutions. In fact, the associated characteristic equation will be a *quadratic* equation (Eq. 4.5 with $N = 2$). A quadratic equation has exactly two solutions. If these solutions are complex, they must occur in a conjugate pair, that is, $\alpha \pm i\beta$. Since these are the only solutions of the characteristic equation in this case, Eq. 4.10 must represent the general solution of a *second-order* homogeneous linear differential equation whose associated characteristic equation has complex solutions. This result is summarized as follows.

Case 3(a):

Consider a second-order equation of the form given in Eq. 4.4. That is,

$$b_2\left(\frac{d^2 y(x)}{dx^2}\right) + b_1\left(\frac{dy(x)}{dx}\right) + b_0 y(x) = 0 \qquad (4.11)$$

$$[b_0, b_1, \text{ and } b_2 \text{ are constants.}]$$

The associated characteristic equation (Eq. 4.5 with $N = 2$) is

$$b_2 r^2 + b_1 r + b_0 = 0$$

Suppose that the characteristic equation has two complex (nonreal) conjugate solutions. Then they occur in a conjugate pair. Let these solutions be denoted by

$$r_1 = \alpha + i\beta \quad [\beta \neq 0]$$
$$r_2 = \alpha - i\beta \quad [\beta \neq 0]$$

Since these are the only solutions of the characteristic equation, from Eq. 4.10, Eq. 4.11 has general solution

$$y(x) = e^{\alpha x}(c_1 \cos \beta x + c_2 \sin \beta x) \qquad (4.12)$$

$$[c_1 \text{ and } c_2 \text{ are arbitrary constants.}]$$

Example 4.10

Determine the general solution of the differential equation of Ex. 4.3.

$$\frac{d^2 y}{dx^2} + (2)\left(\frac{dy}{dx}\right) + 2y = 0$$

Solution:

From Ex. 4.3, the associated characteristic equation is given by

$$r^2 + 2r + 2 = 0$$

The solutions of this equation were found in Ex. 4.3.

$$r = \frac{-2 \pm \sqrt{4 - (4)(2)}}{2}$$

$$= \frac{-2 \pm \sqrt{-4}}{2}$$

$$= \frac{-2 \pm \sqrt{4i^2}}{2}$$

Simplifying,

$$r = -1 \pm i$$

The characteristic equation has complex (nonreal) solutions occurring in a complex conjugate pair (as expected—see Case 3(a)). Since the differential equation is of second order, the general solution is obtained from Eq. 4.12 with $\alpha = -1$ and $\beta = 1$.

$$y(x) = e^{-x}(c_1 \cos x + c_2 \sin x) \qquad \left[\begin{array}{c} c_1 \text{ and } c_2 \text{ are} \\ \text{arbitrary constants.} \end{array}\right]$$

Example 4.11

Determine the general solution of the following differential equation.

$$(2)\left(\frac{d^2 y}{dx^2}\right) + (3)\left(\frac{dy}{dx}\right) + 4y = 0$$

Solution:

First write down the associated characteristic equation.

$$2r^2 + 3r + 4 = 0$$

Using the quadratic formula (Eq. A3.2, App. 3), it can be seen that this equation has the following complex (nonreal) solutions.

$$r = \frac{-3 \pm \sqrt{9 - (4)(2)(4)}}{4}$$

$$= \frac{-3 \pm \sqrt{-23}}{4}$$

$$= \frac{-3 \pm \sqrt{23i^2}}{4}$$

$$= \frac{-3 \pm i\sqrt{23}}{4}$$

$$= -\frac{3}{4} \pm i\frac{\sqrt{23}}{4}$$

Since the differential equation is of second order, the general solution is obtained from Eq. 4.12 with $\alpha = -3/4$ and $\beta = \sqrt{23}/4$.

$$y(x) = e^{-\frac{3}{4}x}\left[c_1 \cos\left(\frac{\sqrt{23}}{4}x\right) + c_2 \sin\left(\frac{\sqrt{23}}{4}x\right)\right]$$

$$[c_1 \text{ and } c_2 \text{ are arbitrary constants.}]$$

The theory developed in Cases 1 through 3(a) can be used to construct general solutions of differential equations of the form of Eq. 4.4, whose associated characteristic equations have a mixture of real, complex, and repeated solutions. In these cases, any of Eqs. 4.8 through 4.10 can be added together, as required. This concept is illustrated in the following examples.

Example 4.12

Solve the following differential equation for y.

$$\frac{d^3 y}{dx^3} + \frac{d^2 y}{dx^2} - 2y = 0$$

Solution:

First write down the associated characteristic equation. As in Eq. 4.5, the characteristic equation is given by

$$r^3 + r^2 - 2 = 0$$

Factor the left-hand side of this equation.

$$(r - 1)(r^2 + 2r + 2) = 0$$

This equation is satisfied by $r = 1$ and $r = -1 \pm i$ (see Ex. 4.11). The characteristic equation has a mixture of real and complex solutions. The general solution of the differential equation is formed by adding Eq. 4.8 with

$N = 1$ and $r_1 = 1$ to Eq. 4.10 (or Eq. 4.12) with $\alpha = -1$ and $\beta = 1$.

$$y(x) = c_1 e^x + e^{-x}(c_2 \cos x + c_3 \sin x)$$

[c_1, c_2, and c_3 are arbitrary constants.]

Example 4.13

Solve the following differential equation for y.

$$\frac{d^3 y}{dx^3} - (3)\left(\frac{dy}{dx}\right) + 2y = 0$$

Solution:

First write down the associated characteristic equation. As in Eq. 4.5, the characteristic equation is given by

$$r^3 - 3r + 2 = 0$$

Factor the left-hand side of this equation.

$$(r - 1)^2(r + 2) = 0$$

This equation is satisfied by $r = 1, 1$ and $r = -2$.

The general solution of the differential equation is formed by adding Eq. 4.8 with $N = 1$ and $r_1 = -2$ to Eq. 4.9 with $r = 1$ and $N = 2$.

$$y(x) = c_1 e^{-2x} + (c_2 + c_3 x)\, e^x \quad \begin{bmatrix} c_1,\ c_2,\ \text{and}\ c_3\ \text{are} \\ \text{arbitrary constants.} \end{bmatrix}$$

Example 4.14

Solve the following differential equation for y.

$$\frac{d^4 y}{dx^4} - (4)\left(\frac{d^3 y}{dx^3}\right) + (7)\left(\frac{d^2 y}{dx^2}\right) - (6)\left(\frac{dy}{dx}\right) + 2y = 0$$

Solution:

First write down the associated characteristic equation. As in Eq. 4.5, the characteristic equation is given by

$$r^4 - 4r^3 + 7r^2 - 6r + 2 = 0$$

Factor the left-hand side of this equation.

$$(r - 1)^2(r^2 - 2r + 2) = 0$$

This equation is satisfied by $r = 1, 1$ and $r = 1 \pm i$ (as in Ex. 4.10). The characteristic equation has a mixture of real (repeated) and complex solutions. The general solution of the differential equation is formed by adding

Eq. 4.9 with $N = 2$ and $r = 1$ to Eq. 4.10 (or Eq. 4.12) with $\alpha = 1$ and $\beta = 1$.

$$y(x) = (c_1 + c_2 x)e^x + e^x(c_3 \cos x + c_4 \sin x)$$

[c_1, c_2, c_3, and c_4 are arbitrary constants.]

Note that in the previous examples, the number of arbitrary constants in the general solution is the same as the order of the differential equation. This is true for all differential equations in the form of Eq. 4.4 (and, in fact, Eq. 4.1).

First-Order Linear Homogeneous Differential Equations with Constant Coefficients

Consider Eq. 4.4 with $N = 1$.

$$b_1 \frac{dy}{dx} + b_0 y = 0$$

Assuming that $b_1 \neq 0$ ($b_1 = 0$ and $b_0 \neq 0$ mean that this equation has the trivial solution $y(x) = 0$) and recalling the notation $y' = dy/dx$, this equation can be rewritten as

$$y' + ay = 0 \quad [a = b_0/b_1 \text{ is a real constant.}] \quad (4.13)$$

The associated characteristic equation is

$$r + a = 0$$

This equation has the (real) solution $r = -a$. From Eq. 4.8, the general solution of Eq. 4.13 is given by

$$y(x) = ce^{-ax} \quad [c \text{ is an arbitrary constant.}] \quad (4.14)$$

Example 4.15

Solve the following boundary value problem for $y(x)$.

$$y' - y = 0$$
$$y(0) = 10$$

Solution:

First find the general solution of the differential equation; then use the boundary condition to find the arbitrary constant. From Eq. 4.13 (with $a = -1$), the general solution of the differential equation is given by

$$y(x) = ce^x \quad [c \text{ is an arbitrary constant.}]$$

The boundary condition, $y(0) = 10$, is used to find c. In fact, the boundary condition requires that

$$y(0) = ce^0 = 10$$
$$c(1) = 10$$
$$c = 10$$

Finally, the particular, or specific, solution is given by

$$y(x) = 10e^x$$

Example 4.16
Solve the following initial value problem.

$$3\frac{dy(t)}{dt} + 2y(t) = 0$$

$$y(0) = 1$$

Solution:
First find the general solution of the differential equation; then use the initial condition to find the arbitrary constant. Write the differential equation in the standard form given in Eq. 4.13, noting that since the independent variable is t, y' is replaced by dy/dt.

$$\frac{dy}{dt} + \frac{2}{3}y = 0$$

From Eq. 4.14 (with $a = 2/3$), the general solution of the differential equation is given by

$$y(t) = ce^{-\frac{2}{3}t} \quad [c \text{ is an arbitrary constant.}]$$

The initial condition, $y(0) = 1$, requires that

$$y(0) = ce^{-\left(\frac{2}{3}\right)(0)} = 1$$
$$c(1) = 1$$
$$c = 1$$

Finally, the particular, or specific, solution is given by

$$y(t) = e^{-\frac{2}{3}t}$$

Second–Order Linear Homogeneous Differential Equations with Constant Coefficients

Consider Eq. 4.4 with $N = 2$.

$$b_2\frac{d^2y}{dx^2} + b_1\frac{dy}{dx} + b_0y = 0$$

Assuming that $b_2 \neq 0$ ($b_2 = 0$ corresponds to Eq. 4.4 with $N = 1$) and recalling the notation $y'' = d^2y/dx^2$ and $y' = dy/dx$, this equation can be rewritten as

$$y'' + 2ay' + by = 0 \qquad (4.15)$$

In Eq. 4.15, $2a = b_1/b_2$ and $b = b_0/b_2$ are real constants (chosen in this form for convenience).

The theory developed above for Eq. 4.4 allows a complete solution of Eq. 4.15 for any values of the constants a and b. In fact, the associated characteristic equation is

$$r^2 + 2ar + b = 0$$

Using the quadratic formula (Eq. A3.2, App. 3), this equation has the solutions

$$r_{1,2} = \frac{-2a \pm \sqrt{(2a)^2 - 4b}}{2}$$
$$= -a \pm \sqrt{a^2 - b}$$

There are three cases to consider.

Case 1:
If $a^2 > b$, the two solutions $r_{1,2}$ are real and distinct. From Eq. 4.8, the general solution of Eq. 4.15 is, in this case,

$$y(x) = c_1e^{r_1x} + c_2e^{r_2x} \qquad (4.16)$$

In Eq. 4.16, c_1 and c_2 are arbitrary constants. Physically, in this case, Eq. 4.15 models an overdamped system.

Case 2:
If $a^2 = b$, $r_{1,2} = -a$; that is, the two solutions $r_{1,2}$ are real and the same. From Eq. 4.9 with $r = -a$ and $N = 2$, the general solution of Eq. 4.15 is, in this case,

$$y(x) = (c_1 + c_2x)e^{-ax} \qquad (4.17)$$

In Eq. 4.17, c_1 and c_2 are arbitrary constants. Physically, in this case, Eq. 4.15 models a critically damped system.

Case 3:
If $a^2 < b$, the two solutions $r_{1,2}$ are complex (nonreal). In fact,

$$r_{1,2} = -a \pm i\sqrt{b - a^2}$$

From Eq. 4.12 with $\alpha = -a$ and $\beta = \sqrt{b - a^2}$, the general solution of Eq. 4.15 is, in this case,

$$y(x) = e^{\alpha x}(c_1 \cos \beta x + c_2 \sin \beta x)$$
$$= e^{-ax}\left[c_1 \cos\left(x\sqrt{b - a^2}\right) + c_2 \sin\left(x\sqrt{b - a^2}\right)\right]$$

$$[c_1 \text{ and } c_2 \text{ are arbitrary constants.}] \qquad (4.18)$$

Physically, in this case, Eq. 4.15 models an underdamped system.

Example 4.17
Solve the following boundary value problem for $y(x)$.

$$y'' + 6y' + 10y = 0$$
$$y(0) = 0$$
$$y'(0) = 1$$

Solution:

First find the general solution of the differential equation; then use the boundary conditions to find the constants. There are two ways to find the general solution of the differential equation. The first way involves a comparison of the differential equation with Eq. 4.15. In fact, the differential equation is basically Eq. 4.15 with $a = 3$ and $b = 10$. Noting that $a^2 < b$ and $b - a^2 = 1$, from Eq. 4.18 the general solution of the differential equation is given by

$$y(x) = e^{-3x}(c_1 \cos x + c_2 \sin x) \quad \left[\begin{array}{c} c_1 \text{ and } c_2 \text{ are} \\ \text{arbitrary constants.} \end{array}\right]$$

This general solution can also be obtained by proceeding as in Exs. 4.11 and 4.12.

In fact, the characteristic equation is given by

$$r^2 + 6r + 10 = 0$$

This equation has complex (nonreal) solutions given by $r = -3 \pm i$. The general solution is now obtained from Eq. 4.12.

The boundary condition $y(0) = 0$ requires that

$$y(0) = e^{-(3)(0)}(c_1 \cos 0 + c_2 \sin 0) = 0$$
$$(1)[c_1(1) + c_2(0)] = 0$$
$$c_1 = 0$$

To apply the condition $y'(0) = 1$, first find $y'(x)$ from the general solution $y(x)$ (noting that $c_1 = 0$). Use the product rule for differentiation (Eq. 1.5).

$$y(x) = e^{-3x}(c_1 \cos x + c_2 \sin x) = c_2 e^{-3x} \sin x$$
$$y'(x) = c_2 \frac{d}{dx}(e^{-3x} \sin x)$$
$$= c_2(e^{-3x} \cos x - 3e^{-3x} \sin x)$$

The condition $y'(0) = 1$ requires that

$$c_2[e^{-(3)(0)} \cos 0 - 3e^{-(3)(0)} \sin(0)] = 1$$
$$c_2[1 - (3)(0)] = 1$$
$$c_2 = 1$$

Finally, the particular, or specific, solution is

$$y(x) = e^{-3x}(c_1 \cos x + c_2 \sin x)$$
$$= e^{-3x}[(0)(\cos x) + (1)(\sin x)]$$
$$= e^{-3x} \sin x$$

Example 4.18

Solve the following initial value problem for $y(t)$.

$$\frac{d^2y}{dt^2} + 6\frac{dy}{dt} + 9y = 0$$
$$y(0) = 1$$
$$\frac{dy}{dt}(0) = 0$$

Solution:

First find the general solution of the differential equation; then use the initial conditions to evaluate the constants. As in Ex. 4.17, there are two ways to find the general solution of the differential equation. First proceed as in Ex. 4.7. The characteristic equation is given by

$$r^2 + 6r + 9 = 0$$

Factor the left-hand side of this equation.

$$(r + 3)^2 = 0$$

This equation is satisfied by $r = -3, -3$. Consequently, the characteristic equation has a real, repeated solution. By Eq. 4.9, the general solution of the differential equation is given by

$$y(t) = (c_1 + c_2 t)e^{-3t} \quad \left[\begin{array}{c} c_1 \text{ and } c_2 \text{ are} \\ \text{arbitrary constants.} \end{array}\right]$$

Alternatively, compare the differential equation with Eq. 4.15 and note that $a = 3$ and $b = 9$. Consequently, $a^2 = b$ and Eq. 4.17 (with $a = 3$) again leads to the general solution $y(t) = (c_1 + c_2 t)e^{-3t}$. Use the initial conditions to evaluate the constants. The condition $y(0) = 1$ requires

$$y(0) = [c_1 + c_2(0)]e^{-(3)(0)} = 1$$
$$(c_1 + 0)(1) = 1$$
$$c_1 = 1$$

To apply the second condition $(dy/dt)(0) = 0$, find dy/dt from the expression for $y(t)$ but use the fact that $c_1 = 1$.

$$y(t) = (1 + c_2 t)e^{-3t}$$
$$\frac{dy}{dt} = \frac{d}{dt}[(1 + c_2 t)e^{-3t}]$$
$$= \frac{d}{dt}(e^{-3t}) + c_2 \frac{d}{dt}(te^{-3t})$$
$$= -3e^{-3t} + c_2(e^{-3t} - 3te^{-3t})$$

The condition $(dy/dt)(0) = 0$ requires

$$-3e^{-(3)(0)} + c_2[e^{-(3)(0)} - (3)(0)e^{-(3)(0)}] = 0$$
$$-(3)(1) + c_2[1 - (3)(0)] = 0$$

Solving this equation leads to $c_2 = 3$. Finally, the particular, or specific, solution is given by

$$y(t) = (c_1 + c_2 t)e^{-3t}$$
$$= (1 + 3t)e^{-3t}$$

To note the significance of the results developed in this section, recall the differential equation in Eq. 4.1 but assume that the coefficients $b_0, b_1, \ldots b_N$ are constant. That is,

$$b_N \frac{d^N y(x)}{dx^N} + \cdots + b_1 \frac{dy(x)}{dx} + b_0 y(x) = f(x) \qquad \textbf{(4.19)}$$

$$[b_0, b_1, \ldots b_N \text{ are constants.}]$$

In Eq. 4.19, the right-hand side, $f(x)$, is some known (forcing) function.

The general solution, $y(x)$, of this differential equation can be decomposed into two parts. According to Eq. 4.2,

$$y(x) = y_h(x) + y_p(x) \qquad \textbf{(4.20)}$$

The complementary solution, $y_h(x)$, is the general solution of Eq. 4.3. With the coefficients $b_0, b_1, \ldots b_N$ assumed to be constant, Eq. 4.3 is basically Eq. 4.4. That is, $y_h(x)$ is the general solution of Eq. 4.4. Equation 4.4 has been studied extensively in this section. Consequently, the theory developed in this section can be used to find the complementary solution, $y_h(x)$, for many different differential equations of the form of Eq. 4.19.

Example 4.19
Find the complementary solution for the nonhomogeneous differential equation $y'' + 6y' + 10y = e^x \sin x$.

Solution:
The complementary solution $y_h(x)$ is the general solution of the following homogeneous differential equation.

$$y_h'' + 6y_h' + 10y_h = 0$$

From Ex. 4.17, this equation has the following general solution.

$$y_h(x) = e^{-3x}(c_1 \cos x + c_2 \sin x) \qquad \left[\begin{array}{c} c_1 \text{ and } c_2 \text{ are} \\ \text{arbitrary constants.} \end{array}\right]$$

Therefore, the complementary solution of the nonhomogeneous equation $y'' + 6y' + 10y = e^x \sin x$ is

$$y_h(x) = e^{-3x}(c_1 \cos x + c_2 \sin x) \qquad \left[\begin{array}{c} c_1 \text{ and } c_2 \text{ are} \\ \text{arbitrary constants.} \end{array}\right]$$

Example 4.20
Find the complementary solution for the differential equation $y'' - 2y' - 3y = 3e^{4x}$.

Solution:
The complementary solution $y_h(x)$ is the general solution of the following homogeneous differential equation.

$$y_h'' - 2y_h' - 3y_h = 0$$

From Ex. 4.4, this equation has the following general solution.

$$y_h(x) = c_1 e^{-x} + c_2 e^{3x} \qquad \left[\begin{array}{c} c_1 \text{ and } c_2 \text{ are} \\ \text{arbitrary constants.} \end{array}\right]$$

Therefore, the complementary solution for the nonhomogeneous equation $y'' - 2y' - 3y = 3e^{4x}$ is

$$y_h(x) = c_1 e^{-x} + c_2 e^{3x} \qquad \left[\begin{array}{c} c_1 \text{ and } c_2 \text{ are} \\ \text{arbitrary constants.} \end{array}\right]$$

To complete the solution of differential equations of the form of Eq. 4.19, it is necessary to develop methods for finding the particular solutions $y_p(x)$. (See Eq. 4.20 and the discussion surrounding Eq. 4.2.) These methods are the subject of the next section.

PRACTICE PROBLEMS
Find the general solution of the differential equations in Probs. 1 through 4. In each case, y is a function of x.

1. (a) $y' + 3y = 0$
   (b) $y' - \frac{1}{2}y = 0$
   (c) $5y' - 2y = 0$

2. (a) $y'' - y' = 0$
   (b) $y'' - 9y = 0$
   (c) $y'' - 2y' - 3y = 0$

3. (a) $y'' - 16y' + 64y = 0$
   (b) $5y'' + 50y' + 125y = 0$

4. (a) $y'' + y' + y = 0$
   (b) $y'' - 4y' + 5y = 0$
   (c) $y'' + 16y = 0$

5. Solve the following initial value problem for $y(t)$.

$$\frac{dy}{dt} - 5y = 0$$
$$y(0) = 3$$

6. Solve the following boundary value problem for $y(x)$.

$$y'' + 4y = 0$$
$$y(0) = 0$$
$$y\left(\frac{\pi}{4}\right) = 1$$

7. Determine the general solution of the following differential equation.

$$\frac{d^5y}{dx^5} + \frac{d^4y}{dx^4} - \frac{d^3y}{dx^3} - (3)\left(\frac{d^2y}{dx^2}\right) + 2y = 0$$

Hint: The characteristic equation is given by

$$r^5 + r^4 - r^3 - 3r^2 + 2 = (r+1)(r^2+2r+2)(r-1)^2 = 0$$

2 Linear Nonhomogenous Differential Equations with Constant Coefficients

There are many different ways of obtaining a particular solution, $y_p(x)$, of Eq. 4.1. When the coefficients $b_0, b_1, \ldots b_N$ are constant (see Eq. 4.19), one of the most systematic ways of obtaining $y_p(x)$ is known as the *method of undetermined coefficients*. This method uses the form of the forcing function, $f(x)$, on the right-hand side of the differential equation to suggest a form for $y_p(x)$. This "suggestion" for $y_p(x)$ contains undetermined constants (the *undetermined coefficients*), which are evaluated by substituting the expression for $y_p(x)$ back into the differential equation.

Example 4.21
Find a particular solution, $y_p(x)$, of the following nonhomogeneous differential equation.

$$y'' + y' + y = 5$$

Solution:
The forcing function (comparing with Eq. 4.19) is given by $f(x) = 5$. This suggests that a particular solution, $y_p(x)$, will take the same form. Let $y_p(x) = A =$ constant and substitute y_p into the differential equation to find the specific value of A.

$$y_p'' + y_p' + y_p = 5$$
$$\frac{d^2}{dx^2}(A) + \frac{d}{dx}(A) + A = 5$$
$$0 + 0 + A = 5$$
$$A = 5$$

It follows that $y_p(x) = A = 5$ is a particular solution of the differential equation. To check this result, substitute $y_p(x) = 5$ back into the left-hand side of the nonhomogeneous differential equation. Since y_p is a particular

solution, the right-hand side of the differential equation (i.e., 5) should result.

$$\frac{d^2}{dx^2}(5) + \frac{d}{dx}(5) + 5 = 0 + 0 + 5 = 5$$

Example 4.22
Find a particular solution, $y_p(x)$, of the following nonhomogeneous differential equation.

$$y'' + 2y' + y = 2e^x$$

Solution:
The right-hand side of the differential equation takes the form of an exponential function. This suggests that a particular solution, $y_p(x)$, will also take the form of an exponential function. Let $y_p(x) = Ae^x$ (where A is constant), and substitute into the differential equation to find the specific value of A.

$$y_p'' + 2y_p' + y_p = 2e^x$$
$$\frac{d^2}{dx^2}(Ae^x) + 2\frac{d}{dx}(Ae^x) + Ae^x = 2e^x$$
$$Ae^x + 2Ae^x + Ae^x = 2e^x$$
$$4Ae^x = 2e^x$$

It follows that $4A = 2$ and $A = 1/2$. Consequently, $y_p(x) = Ae^x = e^x/2$ is a particular solution of the nonhomogeneous differential equation. To check this result, substitute $y_p(x) = e^x/2$ back into the left-hand side of the differential equation. Since y_p is a particular solution, the right-hand side of the differential equation (i.e., $2e^x$) should result.

$$\left(\frac{d^2}{dx^2}\right)\left(\frac{e^x}{2}\right) + 2\frac{d}{dx}\left(\frac{e^x}{2}\right) + \frac{e^x}{2} = \frac{e^x}{2} + 2\frac{e^x}{2} + \frac{e^x}{2}$$
$$= 2e^x$$

Example 4.23
Find a particular solution, $y_p(x)$, of the following nonhomogeneous differential equation.

$$y'' + y' + 2y = x$$

Solution:
The right-hand side of the differential equation takes the form of a first-degree polynomial. This suggests that a particular solution, $y_p(x)$, will also take the form of a first-degree polynomial. Let $y_p(x) = Ax + B$, where A and B are constants. Substitute this $y_p(x)$ into

the differential equation to find the specific values of A and B.

$$y_p'' + y_p' + 2y_p = x$$

$$\frac{d^2}{dx^2}(Ax + B) + \frac{d}{dx}(Ax + B) + (2)(Ax + B) = x$$

$$0 + A + (2)(Ax + B) = x$$

$$2Ax + (A + 2B) = x$$

Compare coefficients of terms with different powers of x.

$$2A = 1$$
$$A + 2B = 0$$

Solving these equations by elimination leads to $A = 1/2$ and $B = -1/4$. It follows that $y_p(x) = Ax + B = (1/2)x - (1/4)$ is a particular solution of the nonhomogeneous differential equation. To check this result, substitute $y_p(x) = (x/2) - (1/4)$ back into the left-hand side of the differential equation. The right-hand side of the differential equation (i.e., x) should result.

$$\frac{d^2}{dx^2}\left(\frac{x}{2} - \frac{1}{4}\right) + \frac{d}{dx}\left(\frac{x}{2} - \frac{1}{4}\right) + (2)\left(\frac{x}{2} - \frac{1}{4}\right)$$

$$= 0 + \frac{1}{2} + \left(x - \frac{1}{2}\right) = x$$

The method of undetermined coefficients (illustrated in Exs. 4.21 through 4.23), as presented here, can be used only when the forcing function, $f(x)$, on the right-hand side of the nonhomogeneous differential equation (see Eq. 4.19) takes on one of the specific forms in Table 4.1. In each of these cases, the form of the corresponding particular solution can be read from Table 4.1. For any other form of $f(x)$, the particular solution must be found using some other technique.

Table 4.1 Particular Solutions

form of $f(x)$	form of $y_p(x)$
A	$x^s B$
$A e^{\alpha x}$	$x^s B e^{\alpha x}$
$A_1 \sin \omega x + A_2 \cos \omega x$	$x^s(B_1 \sin \omega x + B_2 \cos \omega x)$
$P_n(x)$	$x^s Q_n(x)$
$P_n(x)e^{\alpha x}$	$x^s Q_n(x)e^{\alpha x}$
$P_n(x)e^{\alpha x} \sin \omega x$	$x^s[Q_n(x)e^{\alpha x}\cos \omega x + R_n(x)e^{\alpha x}\sin \omega x]$
$P_n(x)e^{\alpha x} \cos \omega x$	$x^s[Q_n(x)e^{\alpha x}\cos \omega x + R_n(x)e^{\alpha x}\sin \omega x]$

In Table 4.1, A is a given (non-zero) constant; A_1 and A_2 are given constants (not both zero); n, α, and ω

are given constants; $P_n(x) = a_n x^n + a_{n-1}x^{n-1} + \cdots + a_0$ where $a_0, \ldots, a_{n-1}, a_n$ are given constants (some of which may be zero); B is a specific (non-zero) constant to be determined; B_1 and B_2 are specific constants to be determined (not both zero); $Q_n(x) = b_n x^n + b_{n-1}x^{n-1} + \cdots + b_0$ where $b_0, \ldots, b_{n-1}, b_n$ are specific constants to be determined; and R_n is the polynomial defined by $R_n(x) = C_n x^n + C_{n-1}x^{n-1} + \cdots + C_0$, where $C_0, \ldots, C_{n-1}, C_n$ are specific constants to be determined.

The exponent s in Table 4.1 is the smallest positive integer $(0, 1, 2, \ldots)$ required to ensure that no term in the suggested particular solution $y_p(x)$ is also a solution to the corresponding homogeneous differential equation. That is, s is the lowest positive integer required to ensure that no term in $y_p(x)$ also appears in $y_h(x)$. The required value of s is determined before the coefficients in $y_p(x)$ are determined (see Exs. 4.24 through 4.38).

The following examples illustrate the use of Table 4.1 in finding particular solutions of nonhomogeneous differential equations of the form of Eq. 4.19. Equation 4.20 is then used to find the complete or general solution of the nonhomogeneous differential equation.

Example 4.24
Find the general, or complete, solution of the following differential equation.

$$y'' + 3y' + 2y = 3e^{-5x} \qquad \text{(Eq. A)}$$

Solution:
From Eq. 4.20, the general solution is the sum of the complementary solution, $y_h(x)$, and a particular solution, $y_p(x)$. Consider each separately.

Complementary solution:
The complementary solution, $y_h(x)$, is the general solution of the corresponding homogeneous equation. The homogeneous equation is

$$y'' + 3y' + 2y = 0 \qquad \text{(Eq. B)}$$

The characteristic equation is

$$r^2 + 3r + 2 = 0$$

Factor the left-hand side of this equation.

$$(r + 1)(r + 2) = 0$$

The solutions are $r_{1,2} = -2, -1$. These solutions are real and distinct. From Eq. 4.8 with $N = 2$, $r_1 = -1$, and $r_2 = -2$ (or from Eq. 4.16), the general solution of the homogeneous equation is

$$y(x) = c_1 e^{-x} + c_2 e^{-2x} \qquad \left[\begin{array}{l}c_1 \text{ and } c_2 \text{ are} \\ \text{arbitrary constants.}\end{array}\right]$$

Consequently, the complementary solution is given by

$$y_h(x) = c_1 e^{-x} + c_2 e^{-2x} \qquad \text{(Eq. C)}$$

[c_1 and c_2 are arbitrary constants.]

Particular solution:
The forcing function, $f(x)$, on the right-hand side of the nonhomogeneous differential equation (Eq. A) is of the form Ae^{-5x} where, in this case, $A = 3$. According to Table 4.1, suggest a particular solution, $y_p(x)$, of the same form.

$$y_p(x) = x^s B e^{-5x}$$

Here, B is a (non-zero) constant to be determined. To determine the required value of s, start with the lowest possible value of s, that is, $s = 0$. Next, check to see if $y_p(x) = x^0 B e^{-5x} = B e^{-5x}$ solves the corresponding homogeneous equation (Eq. B). There are two ways to check this. The first is perhaps the more tedious of the two and involves direct substitution of $y_p(x) = B e^{-5x}$ into the left-hand side of the homogeneous equation, Eq. B.

$$\frac{d^2}{dx^2}(Be^{-5x}) + 3\frac{d}{dx}(Be^{-5x}) + (2)(Be^{-5x})$$
$$= 25Be^{-5x} - 15Be^{-5x} + 2Be^{-5x}$$
$$= 12Be^{-5x} \neq 0$$

Since the result is not zero, $y_p(x) = Be^{-5x}$ does not solve the homogeneous equation.

The second method is based on an examination of the complementary solution, $y_h(x)$ (Eq. C). This method is quicker and involves less computation. Recall that as the general solution of the corresponding homogeneous equation, $y_h(x)$ "contains," or describes, all solutions of the homogeneous equation. Consequently, every solution of the homogeneous equation can be obtained from y_h by a particular choice of the constants c_1 and c_2. However, in the case of the expression $y_p(x) = Be^{-5x}$, no values of the constants c_1 and c_2 will lead to a solution of the homogeneous equation of the form Be^{-5x}. It follows (from either method) that $y_p(x) = Be^{-5x}$ does not solve the corresponding homogeneous equation. Consequently, 0 is the correct value of s, and $y_p(x) = Be^{-5x}$ is the correct form for the particular solution.

To determine the value of the constant B, find y_p' and y_p''.

$$y_p' = -5Be^{-5x}$$
$$y_p'' = 25Be^{-5x}$$

Substitute Eq. A into the nonhomogeneous equation.

$$y_p'' + 3y_p' + 2y_p = 3e^{-5x}$$
$$25Be^{-5x} - 15Be^{-5x} + 2Be^{-5x} = 3e^{-5x}$$
$$12Be^{-5x} = 3e^{-5x}$$

Comparing coefficients of e^{-5x} on both sides of this equation leads to

$$12B = 3$$

This equation has solution $B = 1/4$. The particular solution is therefore

$$y_p(x) = \left(\frac{1}{4}\right)e^{-5x}$$

The general or complete solution of the nonhomogeneous equation is

$$y(x) = y_h(x) + y_p(x)$$
$$= c_1 e^{-x} + c_2 e^{-2x} + \left(\frac{1}{4}\right)e^{-5x}$$

Example 4.25
Solve the following differential equation.

$$y'' + y' = 3 \qquad \text{(Eq. A)}$$

Solution:
From Eq. 4.20, the general solution is the sum of the complementary solution, $y_h(x)$, and a particular solution, $y_p(x)$. Consider each separately.

Complementary solution:
The complementary solution, $y_h(x)$, is the general solution of the corresponding homogeneous equation. The homogeneous equation is

$$y'' + y' = 0 \qquad \text{(Eq. B)}$$

The characteristic equation is

$$r^2 + r = 0$$

Factor the left-hand side of this equation.

$$r(r + 1) = 0$$

The solutions are $r_{1,2} = -1, 0$. These solutions are real and distinct. From Eq. 4.8 with $N = 2$, $r_1 = 0$, and $r_2 = -1$ (or Eq. 4.16), the general solution of the homogeneous equation is

$$y(x) = c_1 + c_2 e^{-x} \quad [c_1 \text{ and } c_2 \text{ are arbitrary constants.}]$$

Consequently, the complementary solution is given by

$$y_h(x) = c_1 + c_2 e^{-x} \qquad \begin{bmatrix} c_1 \text{ and } c_2 \text{ are} \\ \text{arbitrary constants.} \end{bmatrix} \quad \text{(Eq. C)}$$

Particular solution:

The forcing function, $f(x)$, on the right-hand side of the nonhomogeneous differential equation (Eq. A) is of the form A where, in this case, $A = 3$. According to Table 4.1, suggest a particular solution, $y_p(x)$, of the same form.

$$y_p(x) = x^s B$$

Here, B is a (non-zero) constant to be determined. To determine the required value of s, start with the lowest possible value of s, that is, $s = 0$. Next, check to see if $y_p(x) = x^0 B = B$ solves the corresponding homogeneous equation, Eq. B. Again, there are two ways to do this. The first involves substituting $y_p(x) = B$ into the left-hand side of Eq. B.

$$\frac{d^2}{dx^2}(B) + \frac{d}{dx}(B) = 0 + 0 = 0$$

Since the result is zero for any value of the constant B, $y_p(x) = B$ does solve the homogeneous equation.

This result is confirmed by examining the complementary solution, $y_h(x)$, in Eq. C. A solution of the form $y(x) = B$ will result when the constants c_1 and c_2 take the values $c_1 = B$ and $c_2 = 0$. Consequently, using either method, $y_p(x) = B$ does solve the corresponding homogeneous equation.

It follows that 0 is not the correct value of s. Try the next value of s, that is, $s = 1$. Follow the same procedure and check to see if $y_p(x) = x^1 B = xB$ satisfies the homogeneous equation. In fact, no value of the constants c_1 and c_2 in Eq. C will lead to a solution of the form $y_p(x) = xB$. It follows that $y_p(x) = xB$ does not solve the corresponding homogeneous equation. (This can be verified by substituting $y = xB$ into the left-hand side of Eq. B to obtain a non-zero result). Consequently, 1 is the correct value of s, and $y_p(x) = xB$ is the correct form of the particular solution.

To determine the value of the constant B, find y_p' and y_p''.

$$y_p' = B$$
$$y_p'' = 0$$

Substitute into the nonhomogeneous equation, Eq. A.

$$y_p'' + y_p' = 3$$
$$0 + B = 3$$
$$B = 3$$

The particular solution is therefore

$$y_p(x) = 3x$$

The general, or complete, solution of the nonhomogeneous equation is

$$y(x) = y_h(x) + y_p(x)$$
$$= c_1 + c_2 e^{-x} + 3x$$

Note that a particular solution of the form $y_p(x) = B$ (i.e., with the incorrect value of s) would fail since $y_p(x) = B$ cannot satisfy the nonhomogeneous equation for any value of B. In fact, for $y_p(x) = B$,

$$y_p'' + y_p' = 3$$
$$0 + 0 = 3$$
$$0 = 3$$

(Of course, this last equation is impossible.)

Example 4.26

Solve the following differential equation.

$$4y'' - 4y' + y = 2x$$

Solution:

The complete solution is the sum of the complementary and particular solutions.

Complementary solution:

The homogeneous equation is

$$4y'' - 4y' + y = 0$$

The characteristic equation is

$$4r^2 - 4r + 1 = 0$$

The solutions are

$$r_{1,2} = \frac{-(-4) \pm \sqrt{(-4)^2 - (4)(4)(1)}}{(2)(4)}$$
$$= \frac{4 \pm 0}{8} = \frac{1}{2}, \frac{1}{2}$$

From Eq. 4.17 (with $a = -1/2$), the complementary solution is given by

$$y_h(x) = (c_1 + c_2 x)e^{\frac{1}{2}x} \quad \left[\begin{array}{l} c_1 \text{ and } c_2 \text{ are} \\ \text{arbitrary constants.} \end{array}\right]$$

Particular solution:

The nonhomogeneous forcing function takes the form of a first-degree polynomial. Using Table 4.1, assume the particular solution takes the same form, $y_p(x) = x^s(Ax + B)$. Choose the lowest value of s so that no term in y_p is part of y_h. Start with $s = 0$. Then,

$$y_p(x) = x^0(Ax + B) = Ax + B$$

Neither of the terms Ax or B in the suggested y_p can be generated from y_h by a choice of the constants c_1 and c_2. Consequently, neither of the terms in the suggested y_p solves the homogeneous equation. (This can be verified by direct substitution of each of Ax and B into the homogeneous equation). It follows that 0 is the correct value of s, and $y_p(x) = Ax + B$ is the correct form for y_p. The first and second derivatives are

$$y_p' = A$$
$$y_p'' = 0$$

Substitute into the nonhomogeneous differential equation to find the constants A and B.

$$4y_p'' - 4y_p' + y_p = 2x$$
$$(4)(0) - (4)A + Ax + B = 2x$$
$$(B - 4A) + Ax = 2x$$

Compare coefficients of terms with different powers of x.

$$B - 4A = 0$$
$$A = 2$$

Finally, $A = 2$ and $B = 8$. Consequently, $y_p(x) = 2x + 8$.

Complete solution:

$$y(x) = y_h(x) + y_p(x)$$
$$= (c_1 + c_2 x)e^{\frac{1}{2}x} + 2x + 8 \quad \begin{bmatrix} c_1 \text{ and } c_2 \text{ are} \\ \text{arbitrary constants.} \end{bmatrix}$$

....

Example 4.27 ....
Solve the differential equation for $y(t)$.

$$\frac{d^2y}{dt^2} + y = \sin 2t$$

Solution:
The complete solution is the sum of the complementary and particular solutions.

Complementary solution:
The homogeneous equation is

$$\frac{d^2y}{dt^2} + y = 0$$

The characteristic equation is

$$r^2 + 1 = 0$$

The solutions are

$$r_{1,2} = \frac{-(0) \pm \sqrt{(0)^2 - (4)(1)(1)}}{(2)(1)}$$
$$= \frac{0 \pm 2i}{2} = \pm i$$

From Eq. 4.18 with $\alpha = 0$ and $\beta = 1$, the complementary solution is given by

$$y_h(t) = c_1 \cos t + c_2 \sin t \quad \begin{bmatrix} c_1 \text{ and } c_2 \text{ are} \\ \text{arbitrary constants.} \end{bmatrix}$$

Particular solution:
The nonhomogeneous forcing function takes the form $\sin 2t$. Using Table 4.1, assume the particular solution takes the following form.

$$y_p(t) = t^s(A \sin 2t + B \cos 2t)$$

Choose the lowest value of s so that neither of the terms in the suggested y_p is part of y_h. Start with $s = 0$. Then the particular solution takes the following form.

$$y_p(t) = t^0(A \sin 2t + B \cos 2t) = A \sin 2t + B \cos 2t$$

In fact, no combination of the constants c_1 and c_2 (in y_h) can generate either of the terms $A \sin 2t$ or $B \cos 2t$ in y_p. Consequently, neither of the terms in the suggested y_p solves the homogeneous equation. (This can be verified by substituting each of $A \sin 2t$ and $B \cos 2t$ into the homogeneous equation.) It follows that 0 is the correct value of s, and $y_p(t) = A \sin 2t + B \cos 2t$ is the correct form for y_p. The first and second derivatives are

$$\frac{dy_p}{dt} = 2A \cos 2t - 2B \sin 2t$$

$$\frac{d^2y_p}{dt^2} = -4A \sin 2t - 4B \cos 2t$$

Substitute into the nonhomogeneous differential equation to find the constants A and B.

$$\frac{d^2y_p}{dt^2} + y_p = \sin 2t$$
$$-4A \sin 2t - 4B \cos 2t + A \sin 2t + B \cos 2t = \sin 2t$$
$$-3A \sin 2t - 3B \cos 2t = \sin 2t$$

Compare coefficients of $\sin 2t$ and $\cos 2t$.

$$-3A = 1$$
$$-3B = 0$$

Finally, $A = -1/3$ and $B = 0$. Consequently, $y_p(t) = -(1/3)\sin 2t$.

Complete solution:

$$y(t) = y_h(t) + y_p(t)$$

$$= c_1 \cos t + c_2 \sin t - \left(\frac{1}{3}\right)\sin 2t$$

[c_1 and c_2 are arbitrary constants.]

Example 4.28

Solve the following differential equation.

$$\frac{d^2y}{dt^2} + y = \cos t$$

Solution:

The complete solution is the sum of the complementary and particular solutions.

Complementary solution:

From Ex. 4.27, the complementary solution is given by

$$y_h(t) = c_1 \cos t + c_2 \sin t \qquad \begin{bmatrix} c_1 \text{ and } c_2 \text{ are} \\ \text{arbitrary constants.} \end{bmatrix}$$

Particular solution:

The nonhomogeneous forcing function takes the form $\cos t$. Using Table 4.1, assume the particular solution takes the following form.

$$y_p(t) = t^s(A\sin t + B\cos t)$$

Choose the lowest value of s so that no term in y_p is part of y_h. Start with $s = 0$ so that

$$y_p(t) = t^0(A\sin t + B\cos t) = A\sin t + B\cos t$$

By choosing the constants $c_1 = B$ and $c_2 = A$ (in y_h), it can be seen that the suggested y_p is part of the complementary solution, y_h. In other words, the suggested y_p (and each term in the suggested y_p) solves the corresponding homogeneous equation. (This can be verified by direct substitution of y_p into the homogeneous equation.) This means that 0 is not the correct value for s. Try $s = 1$. The suggested particular solution is now

$$y_p(t) = t^1(A\sin t + B\cos t) = t(A\sin t + B\cos t)$$

In this case, no term in the suggested y_p can be generated by a choice of the constants c_1 and c_2 in y_h. The presence of the t in the suggested y_p prevents this.

Consequently, 1 is the correct choice of s, and $y_p(t) = t(A\sin t + B\cos t)$ is the correct form for y_p. The first and second derivatives are

$$\frac{dy_p}{dt} = A\sin t + B\cos t + t(A\cos t - B\sin t)$$

$$\frac{d^2y_p}{dt^2} = A\cos t - B\sin t + A\cos t - B\sin t$$
$$+ t(-A\sin t - B\cos t)$$
$$= 2A\cos t - 2B\sin t - tA\sin t - tB\cos t$$

Substitute into the nonhomogeneous differential equation to find the constants A and B.

$$\frac{d^2y_p}{dt^2} + y_p = \cos t$$

$$2A\cos t - 2B\sin t - tA\sin t$$
$$- tB\cos t + t(A\sin t + B\cos t) = \cos t$$

$$2A\cos t - 2B\sin t = \cos t$$

Compare coefficients of $\sin t$ and $\cos t$.

$$2A = 1$$
$$-2B = 0$$

Finally, $A = 1/2$ and $B = 0$. Consequently, $y_p(t) = (1/2)t\sin t$.

Complete solution:

$$y(t) = y_h(t) + y_p(t)$$

$$= c_1 \cos t + c_2 \sin t + \left(\frac{1}{2}\right)t\sin t$$

[c_1 and c_2 are arbitrary constants.]

Example 4.29

Solve the differential equation for $y(x)$.

$$y' + 2y = 4xe^{2x}$$

Solution:

The complete solution is the sum of the complementary and particular solutions.

Complementary solution:

The homogeneous equation is

$$y' + 2y = 0$$

The characteristic equation is

$$r + 2 = 0$$

The solution is

$$r = -2$$

From Eq. 4.14 (with $a = 2$), the complementary solution is given by

$$y_h(x) = ce^{-2x} \quad [c \text{ is an arbitrary constant.}]$$

Particular solution:
The nonhomogeneous forcing function takes the form $4xe^{2x}$. Using Table 4.1, assume the particular solution takes the following form.

$$y_p(x) = x^s(Ax + B)e^{2x}$$

Choose the lowest value of s so that no term in y_p is part of y_h. Start with $s = 0$. Then,

$$y_p(x) = x^0(Ax + B)e^{2x} = (Ax + B)e^{2x}$$

In fact, no value of the constant c in y_h can generate any term in the suggested y_p. Consequently, no term in the suggested y_p solves the homogeneous equation. It follows that 0 is the correct value of s, and $y_p(x) = (Ax + B)\,e^{2x}$ is the correct form for y_p. The first derivative is

$$y_p' = Ae^{2x} + 2Axe^{2x} + 2Be^{2x}$$

Substitute into the nonhomogeneous differential equation to find the constants A and B.

$$y_p' + 2y_p = 4xe^{2x}$$
$$Ae^{2x} + 2Axe^{2x} + 2Be^{2x} + (2)(Ax + B)e^{2x} = 4xe^{2x}$$
$$(A + 4B)e^{2x} + 4Axe^{2x} = 4xe^{2x}$$

Compare coefficients of e^{2x} and xe^{2x}.

$$A + 4B = 0$$
$$4A = 4$$

Finally, $A = 1$ and $B = -1/4$. Consequently, $y_p(x) = [x - (1/4)]e^{2x}$.

Complete solution:

$$y(x) = y_h(x) + y_p(x)$$
$$= ce^{-2x} + \left(x - \frac{1}{4}\right)e^{2x} \quad \left[\begin{array}{c} c \text{ is an} \\ \text{arbitrary constant.} \end{array}\right]$$

Example 4.30
Solve the differential equation for $y(x)$.

$$y' - y = 2xe^x$$

Solution:
The complete solution is the sum of the complementary and particular solutions.

Complementary solution:
The homogeneous equation is

$$y' - y = 0$$

The characteristic equation is

$$r - 1 = 0$$

The solution is

$$r = 1$$

From Eq. 4.14 (with $a = -1$), the complementary solution is given by

$$y_h(x) = ce^x \quad [c \text{ is an arbitrary constant.}]$$

Particular solution:
The nonhomogeneous forcing function takes the form $2xe^x$. Using Table 4.1, assume the particular solution takes the following form.

$$y_p(x) = x^s(Ax + B)e^x$$

Choose the lowest value of s so that no term in y_p is part of y_h. Start with $s = 0$. Then,

$$y_p(x) = x^0(Ax + B)e^x = (Ax + B)\,e^x$$

Choosing the value of the constant c in y_h to be B, it can be seen that the term Be^x in the suggested y_p is part of y_h. Consequently, the term Be^x solves the homogeneous equation. (This can be verified by direct substitution of Be^x into the homogeneous equation.) It follows that 0 is not the correct value of s. Try $s = 1$. Then the suggested y_p becomes

$$y_p(x) = x^1(Ax + B)e^x = x\,(Ax + B)\,e^x$$

This time, no value of the constant c in y_h will generate any term in y_p. Consequently, no term in y_p solves the homogeneous equation. It follows that 1 is the correct choice of s, and $y_p(x) = x(Ax+B)e^x$ is the correct form for y_p. The first derivative is

$$y_p' = 2Axe^x + Ax^2e^x + xBe^x + Be^x$$
$$= (2A + B)xe^x + Ax^2e^x + Be^x$$

Substitute into the nonhomogeneous differential equation to find the constants A and B.

$$y_p' - y_p = 2xe^x$$
$$(2A + B)xe^x + Ax^2e^x + Be^x - x(Ax + B)e^x = 2xe^x$$
$$2xe^x A + Be^x = 2xe^x$$

Compare coefficients of e^x and xe^x.

$$B = 0$$
$$2A = 2$$

Finally, $A = 1$ and $B = 0$. Consequently, $y_p(x) = x^2e^x$.

Complete solution:

$$y(x) = y_h(x) + y_p(x)$$
$$= ce^x + x^2e^x \quad [c \text{ is an arbitrary constant.}]$$

Example 4.31

Solve the differential equation for $y(t)$.

$$\frac{d^2y}{dt^2} + 2\frac{dy}{dt} + 2y = e^t \cos 2t$$

Solution:

The complete solution is the sum of the complementary and particular solutions.

Complementary solution:

The homogeneous equation is

$$\frac{d^2y}{dt^2} + 2\frac{dy}{dt} + 2y = 0$$

The characteristic equation is

$$r^2 + 2r + 2 = 0$$

The solutions are

$$r_{1,2} = \frac{-(2) \pm \sqrt{(4)^2 - (4)(1)(2)}}{(2)(1)}$$
$$= \frac{-2 \pm i2\sqrt{2}}{2} = -1 \pm i\sqrt{2}$$

From Eq. 4.18 (with $\alpha = -1$ and $\beta = \sqrt{2}$), the complementary solution is given by

$$y_h(t) = e^{-t}[c_1 \cos(\sqrt{2}t) + c_2 \sin(\sqrt{2}t)]$$
$$[c_1 \text{ and } c_2 \text{ are arbitrary constants.}]$$

Particular solution:

The nonhomogeneous forcing function takes the form $e^t \cos 2t$. Using Table 4.1, assume the particular solution takes the following form.

$$y_p(t) = t^s e^t (A \cos 2t + B \sin 2t)$$

Choose the lowest value of s so that no term in y_p is part of y_h. Start with $s = 0$. Then,

$$y_p(t) = t^0 e^t (A \cos 2t + B \sin 2t) = e^t (A \cos 2t + B \sin 2t)$$

Neither of the terms in the suggested y_p can be generated from y_h by a choice of the constants c_1 and c_2. Consequently, neither of the terms in the suggested y_p solves the homogeneous equation. (This can be verified by direct substitution of each of the terms in y_p into the homogeneous equation.) It follows that 0 is the correct value of s, and $y_p(t) = e^t (A \cos 2t + B \sin 2t)$ is the correct form for y_p. The first and second derivatives are

$$\frac{dy_p}{dt} = e^t (A \cos 2t + B \sin 2t)$$
$$+ e^t (-2A \sin 2t + 2B \cos 2t)$$
$$\frac{d^2y_p}{dt^2} = e^t (A \cos 2t + B \sin 2t)$$
$$+ 2e^t (-2A \sin 2t + 2B \cos 2t)$$
$$+ e^t (-4A \cos 2t - 4B \sin 2t)$$
$$= (A + 4B - 4A)e^t \cos 2t$$
$$+ (B - 4A - 4B)e^t \sin 2t$$

Substitute into the nonhomogeneous differential equation to find the constants A and B.

$$\frac{d^2y_p}{dt^2} + 2\frac{dy_p}{dt} + 2y_p = e^t \cos 2t$$

$$(A + 4B - 4A)e^t \cos 2t$$
$$+ (B - 4A - 4B)e^t \sin 2t$$
$$+ 2e^t[(A + 2B) \cos 2t + (B - 2A) \sin 2t]$$
$$+ 2e^t (A \cos 2t + B \sin 2t) = e^t \cos 2t$$

$$(A + 4B - 4A + 2A + 4B + 2A)e^t \cos 2t$$
$$+ (B - 4A - 4B + 2B - 4A + 2B)e^t \sin 2t = e^t \cos 2t$$

$$(A + 8B)e^t \cos 2t + (B - 8A)e^t \sin 2t = e^t \cos 2t$$

Compare coefficients of $e^t \cos 2t$ and $e^t \sin 2t$.

$$A + 8B = 1$$
$$B - 8A = 0$$

Solving, $A = 1/65$ and $B = 8/65$. Consequently,

$$y_p(t) = \left(\frac{e^t}{65}\right)(\cos 2t + 8\sin 2t)$$

Complete solution:

$$y(t) = y_h(t) + y_p(t)$$
$$= e^{-t}[c_1\cos(\sqrt{2}t) + c_2\sin(\sqrt{2}t)]$$
$$+ \left(\frac{e^t}{65}\right)(\cos 2t + 8\sin 2t)$$

[c_1 and c_2 are arbitrary constants.]

Example 4.32

Solve the differential equation for $y(x)$.

$$y'' + 4y' + 3y = 3xe^x\cos x$$

Solution:

The complete solution is the sum of the complementary and particular solutions.

Complementary solution:

The homogeneous equation is

$$y'' + 4y' + 3y = 0$$

The characteristic equation is

$$r^2 + 4r + 3 = 0$$
$$(r + 3)(r + 1) = 0$$

The solutions are

$$r_{1,2} = -3, -1$$

From Eq. 4.16, the complementary solution is given by

$$y_h(x) = c_1 e^{-3x} + c_2 e^{-x} \quad \begin{bmatrix} c_1 \text{ and } c_2 \text{ are} \\ \text{arbitrary constants.} \end{bmatrix}$$

Particular solution:

The nonhomogeneous forcing function takes the form $P_1(x)e^x\cos x$. Using Table 4.1, assume the particular solution takes the following form.

$$y_p(x) = x^s e^x[(b_1 x + b_0)\cos x + (C_1 x + C_0)\sin x]$$

Choose the lowest value of s so that no term in y_p is part of y_h. Start with $s = 0$. Then,

$$y_p(x) = x^0 e^x[(b_1 x + b_0)\cos x + (C_1 x + C_0)\sin x]$$
$$= e^x[(b_1 x + b_0)\cos x + (C_1 x + C_0)\sin x]$$

None of the terms in the suggested y_p can be generated from y_h by a choice of the constants c_1 and c_2. Consequently, none of the terms in the suggested y_p solves the homogeneous equation. (This can be verified by direct substitution of each of the terms in y_p into the homogeneous equation.) It follows that 0 is the correct value of s, and $y_p(x) = e^x[(b_1 x + b_0)\cos x + (C_1 x + C_0)\sin x]$ is the correct form for y_p. Proceeding as in Ex. 4.30, find the first and second derivatives of y_p and substitute these into the nonhomogeneous equation to find the value of the constants b_1, b_0, C_1, and C_0. In fact,

$$b_1 = \frac{21}{85}$$

$$b_0 = -\frac{738}{7225}$$

$$C_1 = \frac{18}{85}$$

$$C_0 = -\frac{1434}{7225}$$

Consequently,

$$y_p(x) = e^x\left(\left[\left(\frac{21}{85}\right)x - \frac{738}{7225}\right]\cos x + \left[\left(\frac{18}{85}\right)x - \frac{1434}{7225}\right]\sin x\right)$$

Complete solution:

$$y(x) = y_h(x) + y_p(x)$$
$$= c_1 e^{-3x} + c_2 e^{-x} + e^x\left(\left[\left(\frac{21}{85}\right)x - \frac{738}{7225}\right]\cos x + \left[\left(\frac{18}{85}\right)x - \frac{1434}{7225}\right]\sin x\right)$$

[c_1 and c_2 are arbitrary constants.]

Example 4.33

Solve the following boundary value problem for $y(x)$.

$$y'' + 2y' + 5y = 8e^{-x}$$
$$y(0) = 0$$
$$y'(0) = 8$$

Solution:

The complete solution is the sum of the complementary and particular solutions.

Complementary solution:

The homogeneous equation is

$$y'' + 2y' + 5y = 0$$

The characteristic equation is

$$r^2 + 2r + 5 = 0$$

The solutions are

$$r_{1,2} = \frac{-(2) \pm \sqrt{(2)^2 - (4)(1)(5)}}{(2)(1)}$$

$$= \frac{-2 \pm 4i}{2} = -1 \pm 2i$$

From Eq. 4.18 (with $\alpha = -1$ and $\beta = 2$), the complementary solution is given by

$$y_h(x) = e^{-x}(c_1 \cos 2x + c_2 \sin 2x)$$

$$[c_1 \text{ and } c_2 \text{ are arbitrary constants.}]$$

Particular solution:
The nonhomogeneous forcing function takes the form $8e^{-x}$. Using Table 4.1, assume the particular solution takes the same form, $y_p(x) = x^s B e^{-x}$. Choose the lowest value of s so that the suggested y_p is not part of y_h. Start with $s = 0$. Then the particular solution takes the following form.

$$y_p(x) = x^0 B e^{-x} = B e^{-x}$$

In fact, no combination of the constants c_1 and c_2 in y_h can generate the suggested y_p. Consequently, the suggested y_p does not solve the homogeneous equation. (This can be verified by substituting $B e^{-x}$ into the homogeneous equation.) It follows that 0 is the correct value of s, and $y_p(x) = B e^{-x}$ is the correct form for y_p. The first and second derivatives are

$$y_p' = -B e^{-x}$$
$$y_p'' = B e^{-x}$$

Substitute into the nonhomogeneous differential equation to find the constants A and B.

$$y_p'' + 2y_p' + 5y_p = 8e^{-x}$$
$$B e^{-x} - 2B e^{-x} + 5B e^{-x} = 8e^{-x}$$
$$4B e^{-x} = 8e^{-x}$$

Compare coefficients of e^{-x}.

$$4B = 8$$

Finally, $B = 2$. Consequently, $y_p(x) = 2e^{-x}$.

Complete solution:

$$y(x) = y_h(x) + y_p(x)$$
$$= e^{-x}(c_1 \cos 2x + c_2 \sin 2x) + 2e^{-x}$$

$$[c_1 \text{ and } c_2 \text{ are arbitrary constants.}]$$

Use the supplied boundary data to evaluate the constants c_1 and c_2. The condition $y(0) = 0$ requires that

$$y(0) = e^{-0}[c_1 \cos 2(0) + c_2 \sin 2(0)] + 2e^{-0} = 0$$
$$c_1 + 2 = 0$$
$$c_1 = -2$$

The condition $y'(0) = 8$ requires $y'(x)$.

$$y'(x) = -e^{-x}(c_1 \cos 2x + c_2 \sin 2x)$$
$$+ e^{-x}(-2c_1 \sin 2x + 2c_2 \cos 2x) - 2e^{-x}$$

Applying the condition $y'(0) = 8$ leads to

$$-[c_1(1) + c_2(0)] - 2c_1(0) + 2c_2(1) - 2 = 8$$
$$-c_1 + 2c_2 = 10$$

Since $c_1 = -2$, it follows that $c_2 = 4$. Finally, the complete solution (the particular or specific solution of the boundary value problem) becomes

$$y(x) = e^{-x}[(-2)\cos 2x + 4 \sin 2x] + 2e^{-x}$$
$$= -e^{-x}(2 \cos 2x - 4 \sin 2x) + 2e^{-x}$$

Example 4.34
Solve the following initial value problem for $y(t)$.

$$\frac{dy}{dt} = (2)(t - y)$$

$$y(0) = 1$$

Solution:
The complete solution is the sum of the complementary and particular solutions. Write the differential equation in the following form.

$$\frac{dy}{dt} + 2y = 2t$$

Complementary solution:
The homogeneous equation is

$$\frac{dy}{dt} + 2y = 0$$

The characteristic equation is

$$r + 2 = 0$$

The solution is

$$r = -2$$

From Eq. 4.14 (with $a = 2$), the complementary solution is given by

$$y_h(t) = ce^{-2t} \quad [c \text{ is an arbitrary constant.}]$$

Particular solution:
The nonhomogeneous forcing function takes the form of a first-degree polynomial, $2t$. Using Table 4.1, assume the particular solution takes the form $y_p(t) = t^s(At + B)$. Choose the lowest value of s so that no term in y_p is part of y_h. Start with $s = 0$. Then,

$$y_p(t) = t^0(At + B) = At + B$$

In fact, no value of the constant c in y_h can generate either term in the suggested y_p. Consequently, no term in the suggested y_p solves the homogeneous equation. It follows that 0 is the correct value of s, and $y_p(t) = At + B$ is the correct form for y_p. The first derivative is

$$\frac{dy_p}{dt} = A$$

Substitute into the nonhomogeneous differential equation to find the constant A.

$$\frac{dy_p}{dt} + 2y_p = 2t$$
$$A + (2)(At + B) = 2t$$
$$(A + 2B) + 2At = 2t$$

Compare coefficients of terms with different powers of t.

$$A + 2B = 0$$
$$2A = 2$$

Finally, $A = 1$ and $B = -1/2$. Consequently, $y_p(t) = t - (1/2)$.

Complete solution:

$$y(t) = y_h(t) + y_p(t)$$
$$= ce^{-2t} + \left(t - \frac{1}{2}\right) \quad [c \text{ is an arbitrary constant.}]$$

Use the supplied boundary data to evaluate the constant c. The condition $y(0) = 1$ requires that

$$ce^{-(2)(0)} + \left(0 - \frac{1}{2}\right) = 1$$
$$c - \frac{1}{2} = 1$$
$$c = \frac{3}{2}$$

Finally, the complete solution (the particular or specific solution of the initial value problem) becomes

$$y(t) = \left(\frac{3}{2}\right)e^{-2t} + \left(t - \frac{1}{2}\right)$$

Example 4.35
Solve the following initial value problem for $y(t)$.

$$\frac{d^2y}{dt^2} + 16\frac{dy}{dt} + 64y = 8\sin t$$
$$y(0) = 0$$
$$\frac{dy}{dt}(0) = 0$$

Solution:
The complete solution is the sum of the complementary and particular solutions.

Complementary solution:
The homogeneous equation is

$$\frac{d^2y}{dt^2} + 16\frac{dy}{dt} + 64y = 0$$

The characteristic equation is

$$r^2 + 16r + 64 = (r + 8)^2 = 0$$

The solutions are

$$r_{1,2} = -8, -8$$

From Eq. 4.17 (with $a = 8$), the complementary solution is given by

$$y_h(t) = (c_1 + c_2t)e^{-8t} \quad \begin{bmatrix} c_1 \text{ and } c_2 \text{ are} \\ \text{arbitrary constants.} \end{bmatrix}$$

Particular solution:
The nonhomogeneous forcing function takes the form $8\sin t$. Using Table 4.1, assume the particular solution takes the form $y_p(t) = t^s(A\cos t + B\sin t)$. Choose the lowest value of s so that no term in the suggested y_p is part of y_h. Start with $s = 0$. Then the particular solution takes the following form.

$$y_p(t) = t^0(A\cos t + B\sin t) = A\cos t + B\sin t$$

In fact, no combination of the constants c_1 and c_2 (in y_h) can generate either of the terms in the suggested y_p. Consequently, the suggested y_p does not solve the homogeneous equation. (This can be verified by substituting each of $A\cos t$ and $B\sin t$ into the homogeneous equation.) It follows that 0 is the correct value of s and

$y_p(t) = A\cos t + B\sin t$ is the correct form for y_p. The first and second derivatives are

$$\frac{dy_p}{dt} = -A\sin t + B\cos t$$

$$\frac{d^2y_p}{dt^2} = -A\cos t - B\sin t$$

Substitute into the nonhomogeneous differential equation to find the constants A and B.

$$\frac{d^2y_p}{dt^2} + 16\frac{dy_p}{dt} + 64y_p = 8\sin t$$

$$-A\cos t - B\sin t + (16)(-A\sin t + B\cos t)$$
$$+(64)(A\cos t + B\sin t) = 8\sin t$$

$$63A\cos t + 63B\sin t - 16A\sin t + 16B\cos t = 8\sin t$$

$$(63A + 16B)\cos t + (63B - 16A)\sin t = 8\sin t$$

Compare coefficients of $\cos t$ and $\sin t$.

$$63A + 16B = 0$$
$$63B - 16A = 8$$

Solving these equations (by elimination) leads to $A = -(128/4225)$ and $B = 504/4225$. Consequently, $y_p(t) = -(128/4225)\cos t + (504/4225)\sin t$.

Complete solution:

$$y(t) = y_h(t) + y_p(t)$$

$$= (c_1 + c_2 t)e^{-8t} - \left(\frac{128}{4225}\right)\cos t + \left(\frac{504}{4225}\right)\sin t$$

Use the supplied initial data to evaluate the constants c_1 and c_2. The condition $y(0) = 0$ requires that

$$y(0) = [c_1 + c_2(0)]e^{-0} - \left(\frac{128}{4225}\right)\cos 0$$
$$+ \left(\frac{504}{4225}\right)\sin 0$$
$$= 0$$

$$c_1 - \frac{128}{4225} = 0$$

$$c_1 = \frac{128}{4225}$$

The condition $(dy/dt)(0) = 0$ requires dy/dt.

$$\frac{dy}{dt} = c_2 e^{-8t} - (8)(c_1 + c_2 t)e^{-8t} + \left(\frac{128}{4225}\right)\sin t$$
$$+ \left(\frac{504}{4225}\right)\cos t$$

Applying the condition $(dy/dt)(0) = 0$ leads to

$$c_2(1) - 8c_1(1) + \frac{504}{4225} = 0$$

$$c_2 - 8c_1 = -\frac{504}{4225}$$

Since $c_1 = 128/4225$, it follows that

$$c_2 = -\frac{504}{4225} + (8)\left(\frac{128}{4225}\right) = \frac{8}{65}$$

Finally, the complete solution (the particular or specific solution of the boundary value problem) becomes

$$y(t) = \left[\frac{128}{4225} + \left(\frac{8}{65}\right)t\right]e^{-8t} - \left(\frac{128}{4225}\right)\cos t$$
$$+ \left(\frac{504}{4225}\right)\sin t$$

....

Example 4.36

Solve the following boundary value problem for $y(x)$.

$$6y' + y = 8x^2$$
$$y(0) = 1$$

Solution:

The complete solution is the sum of the complementary and particular solutions.

Complementary solution:
The homogeneous equation is

$$6y' + y = 0$$

The characteristic equation is

$$6r + 1 = 0$$

The solution is

$$r = -\frac{1}{6}$$

From Eq. 4.14 (with $a = 1/6$), the complementary solution is given by

$$y_h(x) = ce^{-\frac{1}{6}x} \quad [c \text{ is an arbitrary constant.}]$$

Particular solution:
The nonhomogeneous forcing function takes the form of a second-degree polynomial, $8x^2$. Using Table 4.1,

assume the particular solution takes the form $y_p(x) = x^s(Ax^2 + Bx + E)$. Choose the lowest value of s so that no term in y_p is part of y_h. Start with $s = 0$. Then,

$$y_p(x) = x^0(Ax^2 + Bx + E) = Ax^2 + Bx + E$$

In fact, no value of the constant c in y_h can generate any of the terms in the suggested y_p. Consequently, no term in the suggested y_p solves the homogeneous equation. It follows that 0 is the correct value of s, and $y_p(x) = Ax^2 + Bx + E$ is the correct form for y_p. The first derivative is

$$y_p' = 2Ax + B$$

Substitute into the nonhomogeneous differential equation to find the constants A and B.

$$6y_p' + y_p = 8x^2$$
$$(6)(2Ax + B) + Ax^2 + Bx + E = 8x^2$$
$$Ax^2 + (12A + B)x + 6B + E = 8x^2$$

Compare coefficients of terms with different powers of x.

$$A = 8$$
$$12A + B = 0$$
$$6B + E = 0$$

When these equations are solved by elimination or by Cramer's rule, $A = 8$, $B = -96$, and $E = 576$. Consequently, $y_p(x) = 8x^2 - 96x + 576$.

Complete solution:

$$y(x) = y_h(x) + y_p(x)$$
$$= ce^{-\frac{1}{6}x} + 8x^2 - 96x + 576 \quad \left[\begin{array}{c} c \text{ is an} \\ \text{arbitrary constant.} \end{array}\right]$$

Use the supplied boundary data to evaluate the constant c. The condition $y(0) = 1$ requires that

$$ce^{-\left(\frac{1}{6}\right)(0)} + 576 = 1$$
$$c + 576 = 1$$
$$c = -575$$

Finally, the complete solution (the particular or specific solution of the initial value problem) becomes

$$y(x) = -575e^{-\frac{1}{6}x} + 8x^2 - 96x + 576$$

When the forcing function on the right-hand side of the nonhomogeneous differential equation contains more

than one term of the type appearing in the left-hand column of Table 4.1, Table 4.1 is used to assign a particular solution to each term. The sum of these particular solutions constitutes the final suggestion for a particular solution y_p.

For example, assume the forcing function takes the form $f(x) = f_1(x) + f_2(x)$ where both f_1 and f_2 appear as separate entries in the left-hand column of Table 4.1. Then $y_p(x) = y_{p_1}(x) + y_{p_2}(x)$, with $y_{p_1}(x)$ and $y_{p_2}(x)$ chosen from Table 4.1 according to the respective forms of the functions f_1 and f_2. The procedure is illustrated in Exs. 4.36 and 4.37.

Example 4.37

Find the general or complete solution of the following differential equation.

$$y'' + 9y = 3x - 2e^{-2x}$$

Solution:

The complete solution is the sum of the complementary and particular solutions.

Complementary solution:

The homogeneous equation is

$$y'' + 9y = 0$$

The characteristic equation is

$$r^2 + 9 = 0$$

The solutions are

$$r_{1,2} = \pm 3i$$

From Eq. 4.18 (with $\alpha = 0$ and $\beta = 3$), the complementary solution is given by

$$y_h(x) = c_1\cos 3x + c_2\sin 3x \quad \left[\begin{array}{c} c_1 \text{ and } c_2 \text{ are} \\ \text{arbitrary constants.} \end{array}\right]$$

Particular solution:

The nonhomogeneous forcing function consists of two terms, $f_1(x) = 3x$ and $f_2(x) = -2e^{-2x}$. For each term, use Table 4.1 to identify the corresponding term in a suggested particular solution. From Table 4.1, for the term $f_1(x) = 3x$, suggest $y_{p_1}(x) = x^s(Ax + B)$. From Table 4.1, for the term $f_2(x) = -2e^{-2x}$, suggest $y_{p_2}(x) = x^q E e^{-2x}$. The final suggestion for the particular solution is the sum of $y_{p_1}(x)$ and $y_{p_2}(x)$.

$$y_p(x) = y_{p_1}(x) + y_{p_2}(x) = x^s(Ax + B) + x^q E e^{-2x}$$

The values of s and q are chosen as before. That is, choose the lowest values of s and q so that no term in the suggested y_p is part of y_h. Start with $s = q = 0$. Then the particular solution takes the following form.

$$y_p(x) = Ax + B + Ee^{-2x}$$

In fact, no combination of the constants c_1 and c_2 in y_h can generate any term in the suggested y_p. Consequently, no term in the suggested y_p solves the homogeneous equation. (This can be verified by substituting each term into the homogeneous equation.) It follows that $s = q = 0$ are the correct values of s and q, and $y_p(x) = Ax + B + Ee^{-2x}$ is the correct form for y_p. The first and second derivatives are

$$y_p' = A - 2Ee^{-2x}$$
$$y_p'' = 4Ee^{-2x}$$

Substitute into the nonhomogeneous differential equation to find the constants A and B.

$$y_p'' + 9y_p = 3x - 2e^{-2x}$$
$$4Ee^{-2x} + (9)(Ax + B + Ee^{-2x}) = 3x - 2e^{-2x}$$
$$13Ee^{-2x} + 9Ax + 9B = 3x - 2e^{-2x}$$

Compare coefficients of e^{-2x} and x.

$$13E = -2$$
$$9A = 3$$
$$9B = 0$$

Finally, $A = 1/3$, $B = 0$, and $E = -(2/13)$. Consequently, $y_p(x) = (1/3)x - (2/13)e^{-2x}$.

Complete solution:

$$y(x) = y_h(x) + y_p(x)$$
$$= (c_1 \cos 3x + c_2 \sin 3x) + \left(\frac{1}{3}\right)x - \left(\frac{2}{13}\right)e^{-2x}$$

....

Example 4.38..
Solve the following boundary value problem for $y(x)$.

$$y' + 9y = 2 - 2e^{-9x}$$
$$y(0) = 0$$

Solution:
The complete solution is the sum of the complementary and particular solutions.

Complementary solution:
The homogeneous equation is

$$y' + 9y = 0$$

The characteristic equation is

$$r + 9 = 0$$

The solution is

$$r = -9$$

From Eq. 4.14 (with $a = 9$), the complementary solution is given by

$$y_h(x) = ce^{-9x} \quad [c \text{ is an arbitrary constant.}]$$

Particular solution:
The nonhomogeneous forcing function consists of two terms, $f_1(x) = 2$ and $f_2(x) = -2e^{-9x}$. For each term, use Table 4.1 to identify the corresponding term in a suggested particular solution. From Table 4.1, for the term $f_1(x) = 2$, suggest $y_{p_1}(x) = x^s B$. From Table 4.1, for the term $f_2(x) = -2e^{-9x}$, suggest $y_{p_2}(x) = x^q Ee^{-9x}$. The final suggestion for the particular solution is the sum of $y_{p_1}(x)$ and $y_{p_2}(x)$.

$$y_p(x) = y_{p_1}(x) + y_{p_2}(x) = x^s B + x^q Ee^{-9x}$$

Choose the lowest values of s and q so that no term in y_p is part of y_h. Start with $s = q = 0$. Then,

$$y_p(x) = y_{p_1}(x) + y_{p_2}(x)$$
$$= B + Ee^{-9x}$$

The term $y_{p_1}(x) = B$ cannot be generated from y_h for any value of the constant c. Consequently, the term $y_{p_1}(x) = B$ does not solve the homogeneous equation. (This can be verified by direct substitution of the term B into the homogeneous differential equation.) Zero is the correct value for s. However, the term $y_{p_2}(x) = Ee^{-9x}$ can be generated from y_h by allowing the constant c to take a value of E. Consequently, the term $y_{p_2}(x) = Ee^{-9x}$ solves the homogeneous equation. (This can be verified by direct substitution of the term Ee^{-2x} into the homogeneous differential equation.) It follows that 0 is not the correct value of q for $y_{p_2}(x)$.

Try $q = 1$, but maintain $s = 0$. Then the suggested particular solution takes the following form.

$$y_p(x) = y_{p_1}(x) + y_{p_2}(x)$$
$$= B + xEe^{-9x}$$

In this case, no value of the constant c in y_h can generate any of the terms in the suggested y_{p_1} or y_{p_2}. Consequently, no term in the suggested y_p solves the homogeneous equation. It follows that $s = 0$ and $q = 1$ are the correct values of s and q, and $y_p(x) = B + xEe^{-9x}$ is the correct form for y_p. The first derivative is

$$y_p' = Ee^{-9x} + x(-9Ee^{-9x})$$

Substitute into the nonhomogeneous differential equation to find the constants B and E.

$$y_p' + 9y_p = 2 - 2e^{-9x}$$
$$Ee^{-9x} + x(-9Ee^{-9x}) + 9B + 9xEe^{-9x} = 2 - 2e^{-9x}$$
$$Ee^{-9x} + 9B = 2 - 2e^{-9x}$$

Compare coefficients of the constant term and the term in e^{-9x}.

$$9B = 2$$
$$E = -2$$

When these equations are solved by elimination or by Cramer's rule, $B = 2/9$ and $E = -2$. Consequently,

$$y_p(x) = y_{p_1}(x) + y_{p_2}(x) = \frac{2}{9} - 2xe^{-9x}$$

Complete solution:

$$y(x) = y_h(x) + y_p(x)$$
$$= ce^{-9x} + \frac{2}{9} - 2xe^{-9x} \qquad \left[\begin{array}{c}c \text{ is an}\\ \text{arbitrary constant.}\end{array}\right]$$

Use the supplied boundary data to evaluate the constant c. The condition $y(0) = 0$ requires that

$$ce^{-(9)(0)} + \frac{2}{9} - (2)(0) = 0$$
$$c + \frac{2}{9} = 0$$
$$c = -\frac{2}{9}$$

Finally, the complete solution (the particular or specific solution of the initial value problem) becomes

$$y(x) = -\left(\frac{2}{9}\right)e^{-9x} + \left(\frac{2}{9}\right) - 2xe^{-9x}$$

PRACTICE PROBLEMS

8. Find the general or complete solution of each of the following.

(a) $2y'' + 2y' - 4y = 10e^{3x}$

(b) $4y'' - 4y' + y = \sin x$

(c) $y'' + y' + y = 2x$

(d) $y'' + y' = 4e^{-x}$

9. Solve the following initial value problems.

(a)
$$2\frac{dy}{dt} - 3y = \cos t$$
$$y(0) = 0$$

(b)
$$\frac{dy}{dt} + 8y = 3e^{-8t}$$
$$y(0) = 1$$

(c)
$$\frac{d^2y}{dt^2} + 2\frac{dy}{dt} + y = e^{-t}$$
$$y(0) = 0$$
$$\frac{dy}{dt}(0) = 1$$

10. Solve the following boundary value problem for $y(x)$.

$$y' + 2y = \cos x + \sin x$$
$$y(0) = 0$$

11. Solve the following boundary value problem for $y(x)$.

$$6y'' = 7e^{-2x} + y - y'$$
$$y(0) = 0$$
$$y'(0) = 1$$

FE-Style Exam Problems

In what follows, c_1 and c_2 are arbitrary constants.

1. What is the general solution of the following second-order differential equation?

$$2y'' + 32y = 0$$

(A) $y = c_1 \cos(\sqrt{32}x) + c_2 \sin(\sqrt{32}x)$

(B) $y = c_1 e^{4x} + c_2 e^{-4x}$

(C) $y = c_1 \cos 4x + c_2 \sin 4x$

(D) $y = (c_1 + c_2 x)e^{4x}$

2. What is the solution of the following initial value problem?

$$2\frac{dy}{dt} - 6y = 0$$
$$y(0) = 1$$

(A) $y = e^{3t}$

(B) $y = e^{6t}$

(C) $y = \frac{1}{2}e^{3t}$

(D) $y = e^{-3t}$

3. Solve the following boundary value problem.

$$y'' + 6y' + 9y = 0$$
$$y(0) = 0$$
$$y'(0) = 1$$

(A) $y = (1 + 3x)e^{-x}$
(B) $y = xe^{3x}$
(C) $y = (2 - x)e^{-3x}$
(D) $y = xe^{-3x}$

4. What is the correct general solution for the following differential equation?

$$y'' - 2y' + 5y = 0$$

(A) $y = c_1 \cos 2x + c_2 \sin 2x$
(B) $y = c_1 \cos x + c_2 \sin 2x$
(C) $y = e^x(c_1 \cos 2x + c_2 \sin 2x)$
(D) $y = e^{-x}(c_1 \cos 2x + c_2 \sin 2x)$

5. Which is the correct form of particular solution for the following differential equation? (A and B are definite constants.)

$$\frac{d^2y}{dt^2} + y = \cos t$$

(A) $y_p(t) = A \cos t + B \sin t$
(B) $y_p(t) = t(A \cos t + B \sin t)$
(C) $y_p(t) = A \cos t$
(D) $y_p(t) = At \cos t$

Problems 6–8 refer to the following differential equation and boundary conditions.

$$160y = 36y' - 2y'' + 4e^{3x}$$
$$y(0) = 0$$
$$y'(0) = 1$$

6. Which type of differential equation is shown?
(A) linear, second order, nonhomogeneous
(B) nonlinear, second order, homogeneous
(C) nonlinear, second order, nonhomogeneous
(D) linear, first order, homogeneous

7. Which of the following statements is true for the differential equation?
(A) The equation represents an unstable system.
(B) The complete or general solution is given by the sum of a complementary solution and a particular solution.
(C) The equation may be solved by successive integrations.
(D) The complete solution will contain natural sines and cosines.

8. Which is the correct complete solution of the differential equation with specified boundary conditions?

(A) $y = 3e^{8x} + 7e^{10x} - 1$
(B) $y = -5e^{10x} + 7e^{8x} - \left(\dfrac{2}{37}\right)e^{3x}$
(C) $y = \left(\dfrac{1}{37}\right)(5e^{10x} - 7e^{8x} + 2e^{3x})$
(D) $y = -\left(\dfrac{7}{10}\right)e^{8x} + \left(\dfrac{9}{14}\right)e^{10x} + \left(\dfrac{2}{35}\right)e^{3x}$

9. What is the correct general solution for the following differential equation?

$$y'' - 3y' + 2y = 0$$

(A) $y = c_1 \cos 2x + c_2 \sin 2x$
(B) $y = c_1 e^{-x} + c_2 e^{-2x}$
(C) $y = e^x(c_1 \cos 2x + c_2 \sin 2x)$
(D) $y = c_1 e^x + c_2 e^{2x}$

10. What is the correct general solution for the following differential equation?

$$y'' + 4y' = e^{-4x}$$

(A) $y = c_1 e^{-4x} + c_2 - \left(\dfrac{x}{4}\right)e^{-4x}$
(B) $y = c_1 e^x + c_2 e^{-4x} - \left(\dfrac{x}{4}\right)e^{-4x}$
(C) $y = c_1 e^{-4x} + c_2 - \left(\dfrac{1}{4}\right)e^{-4x}$
(D) $y = c_1 e^x + c_2 e^{2x} - 3$

Appendix

1

Solutions to Practice Problems

Chapter One

1. Use Table 1.1 and Eq. 1.4.

(a)

$$\frac{d}{dx}\left(x^{58} - 50x + \frac{1}{2}\right) = \frac{d}{dx}(x^{58}) - 50\frac{d}{dx}(x) + \frac{d}{dx}\left(\frac{1}{2}\right)$$
$$= 58x^{57} - (50)(1) + 0$$
$$= 58x^{57} - 50$$

(b)

$$\frac{d}{dr}\left(\frac{4}{3}\pi r^3\right) = \left(\frac{4}{3}\right)\pi\frac{d}{dr}(r^3)$$
$$= \left(\frac{4}{3}\right)\pi(3r^2)$$
$$= 4\pi r^2$$

(c)

$$\frac{d}{dt}(\sqrt{10}t^{-\frac{1}{2}}) = (\sqrt{10})\frac{d}{dt}(t^{-\frac{1}{2}})$$
$$= (\sqrt{10})\left(-\frac{1}{2}t^{-\frac{3}{2}}\right)$$
$$= -\left(\frac{\sqrt{10}}{2}\right)t^{-\frac{3}{2}}$$

(d)

$$\frac{d}{dx}(x^{\frac{4}{3}} - x^{\frac{2}{3}}) = \frac{d}{dx}(x^{\frac{4}{3}}) - \frac{d}{dx}(x^{\frac{2}{3}})$$
$$= \left(\frac{4}{3}\right)x^{\frac{1}{3}} - \left(\frac{2}{3}\right)x^{-\frac{1}{3}}$$

2. (a) Use the product rule (Eq. 1.5).

$$\frac{d}{dt}(t^{\frac{1}{3}}(t+2)) = t^{\frac{1}{3}}(1) + (t+2)\left(\frac{1}{3}t^{-\frac{2}{3}}\right)$$
$$= t^{\frac{1}{3}} + \left(\frac{1}{3}\right)t^{-\frac{2}{3}}(t+2)$$

(b) Use the product rule (Eq. 1.5).

$$\frac{d}{dx}((x^2 + x + 1)(x^2 - 2)) = (x^2 + x + 1)(2x)$$
$$+ (x^2 - 2)(2x + 1)$$

(c) Use the quotient rule (Eq. 1.6).

$$\frac{d}{dt}\left(\frac{\sqrt{t} - 1}{\sqrt{t} + 1}\right) = \frac{(\sqrt{t} + 1)\left(\frac{1}{2}t^{-\frac{1}{2}}\right) - (\sqrt{t} - 1)\left(\frac{1}{2}t^{-\frac{1}{2}}\right)}{(\sqrt{t} + 1)^2}$$

This can be factored as follows.

$$\frac{d}{dt}\left(\frac{\sqrt{t} - 1}{\sqrt{t} + 1}\right) = \frac{\frac{1}{2}t^{-\frac{1}{2}}[(\sqrt{t} + 1) - (\sqrt{t} - 1)]}{(\sqrt{t} + 1)^2}$$
$$= \frac{\left(\frac{1}{2}t^{-\frac{1}{2}}\right)(2)}{(\sqrt{t} + 1)^2}$$
$$= \frac{t^{-\frac{1}{2}}}{(\sqrt{t} + 1)^2}$$
$$= \frac{1}{\sqrt{t}(\sqrt{t} + 1)^2}$$

(d) Use the quotient rule (Eq. 1.6).

$$\frac{d}{dx}\left(\frac{1}{x^4 + x^2 - 1}\right) = \frac{(x^4 + x^2 - 1)(0) - (1)(4x^3 + 2x)}{(x^4 + x^2 - 1)^2}$$
$$= -\frac{4x^3 + 2x}{(x^4 + x^2 - 1)^2}$$

3. (a) Use Table 1.1.

$$\frac{d}{dx}(\cos x + 2\tan x) = \frac{d}{dx}(\cos x) + 2\frac{d}{dx}(\tan x)$$
$$= -\sin x + 2\sec^2 x$$

(b) Use the quotient rule (Eq. 1.6).

$$\frac{d}{dx}\left(\frac{\sin x}{1+\cos x}\right) = \frac{(1+\cos x)(\cos x)-(\sin x)(-\sin x)}{(1+\cos x)^2}$$

$$= \frac{\cos x + \cos^2 x + \sin^2 x}{(1+\cos x)^2}$$

Use the trigonometric identity (App. 3) $\cos^2 x + \sin^2 x = 1$.

$$\frac{d}{dx}\left(\frac{\sin x}{1+\cos x}\right) = \frac{\cos x + \cos^2 x + \sin^2 x}{(1+\cos x)^2}$$

$$= \frac{\cos x + 1}{(1+\cos x)^2} = \frac{1}{1+\cos x}$$

(c) Use the product rule (Eq. 1.5).

$$\frac{d}{dx}(x\sin x\cos x) = \frac{d}{dx}[(x\sin x)(\cos x)]$$

$$= (x\sin x)(-\sin x)$$

$$+ (\cos x)\left[\frac{d}{dx}(x\sin x)\right]$$

To find $(d/dx)(x\sin x)$, apply Eq. 1.5 once more.

$$\frac{d}{dx}(x\sin x) = x(\cos x) + (\sin x)(1) = x\cos x + \sin x$$

Finally,

$$\frac{d}{dx}(x\sin x\cos x) = (x\sin x)(-\sin x)$$

$$+ (\cos x)(x\cos x + \sin x)$$

$$= -x\sin^2 x + x\cos^2 x + \cos x\sin x$$

$$= x(\cos^2 x - \sin^2 x) + \cos x\sin x$$

Using trigonometric identities from App. 3, this equation can be rewritten in the following form.

$$\frac{d}{dx}(x\sin x\cos x) = x\cos 2x + \frac{1}{2}\sin 2x$$

Alternatively, since $(1/2)\sin 2x = \sin x\cos x$ (App. 3),

$$\frac{d}{dx}(x\sin x\cos x) = \frac{d}{dx}\left(\frac{x}{2}\sin 2x\right)$$

Use Eq. 1.5 to obtain the derivative as follows.

$$\frac{d}{dx}\left(\frac{x}{2}\sin 2x\right) = \frac{x}{2}(2\cos 2x) + (\sin 2x)\left(\frac{1}{2}\right)$$

$$x\cos 2x + \frac{1}{2}\sin 2x$$

This illustrates how trigonometric identities can be used to simplify expressions before applying the relevant rule for differentiation.

(d) Use the substitution rule (Eq. 1.8).

$$\frac{d}{dx}[\cos(x^3)] = \left[\frac{d}{du}(\cos u)\right]\left(\frac{du}{dx}\right) \quad [u = x^3]$$

$$= (-\sin u)(3x^2)$$

$$= -3x^2\sin(x^3)$$

(e) Use the substitution rule (Eq. 1.8).

$$\frac{d}{dx}[(1+x^{\frac{1}{2}})^{\frac{1}{3}}] = \left[\frac{d}{du}(u^{\frac{1}{3}})\right]\left(\frac{du}{dx}\right) \quad [u = 1+x^{\frac{1}{2}}]$$

$$= \left(\frac{1}{3}u^{-\frac{2}{3}}\right)\left(\frac{1}{2}x^{-\frac{1}{2}}\right)$$

$$= \frac{1}{6}x^{-\frac{1}{2}}(1+x^{\frac{1}{2}})^{-\frac{2}{3}}$$

(f) Use the substitution rule (Eq. 1.8).

$$\frac{d}{dx}(e^{x^2+2}) = \left[\frac{d}{du}(e^u)\right]\left(\frac{du}{dx}\right) \quad [u = x^2 + 2]$$

$$= e^u(2x)$$

$$= 2xe^{x^2+2}$$

4. Use implicit differentiation (Eqs. 1.9 and 1.10).

(a)

$$x^2 = \frac{y^2}{y^2 - 1}$$

Differentiate both sides of this equation with respect to x.

$$2x = \frac{(y^2-1)2yy' - y^2(2yy'-0)}{(y^2-1)^2}$$

$$= \frac{-2yy'}{(y^2-1)^2}$$

Solve for $y' = dy/dx$.

$$y' = \frac{dy}{dx} = -\frac{x(y^2-1)^2}{y}$$

(b)

$$x^2 + y^2 = 5$$

Differentiate both sides of this equation with respect to x.

$$2x + 2yy' = 0$$

Solve for $y' = dy/dx$.

$$y' = \frac{dy}{dx} = -\frac{x}{y}$$

(c)

$$x \sin y + \sin 2y = \cos y$$

Differentiate both sides of this equation with respect to x.

$$x(\cos y)y' + (\sin y)(1) + (2 \cos 2y)y' = (-\sin y)\,y'$$

Solve for $y' = dy/dx$.

$$y'(2 \cos 2y + x \cos y + \sin y = -\sin y$$

$$\begin{aligned} y' &= \frac{dy}{dx} \\ &= \frac{-\sin y}{2 \cos 2y + x \cos y + \sin y} \end{aligned}$$

5. Use Eq. 1.12 with $f(x) = x$ and $g(x) = 2x$.

$$\begin{aligned} \frac{d}{dx}(x^{2x}) &= 2x(x^{2x-1})(1) + x^{2x}(\ln x)(2) \\ &= 2x^{2x} + 2x^{2x} \ln x \\ &= 2x^{2x}(1 + \ln x) \end{aligned}$$

6. (a) The maximum and minimum values occur either at local extrema or at the endpoints. The critical points are located where the first derivative is zero.

$$\begin{aligned} f(x) &= x^3 - 12x + 1 \\ f'(x) &= 3x^2 - 12 \\ 3x^2 - 12 &= 0 \\ (3)(x^2 - 4) &= 0 \\ (3)(x-2)(x+2) &= 0 \\ x &= \pm 2 \end{aligned}$$

Test for a local maximum, local minimum, or inflection point.

$$\begin{aligned} f''(x) &= 6x \\ f''(2) &= (6)(2) = 12 > 0 \quad \text{[local minimum]} \\ f''(-2) &= (6)(-2) = -12 < 0 \quad \text{[local maximum]} \end{aligned}$$

Check the value of the function $f(x)$ at the endpoints, -6 and 4, of the interval, and compare with the value of $f(x)$ at the critical points.

$$\begin{aligned} f(-6) &= (-6)^3 - (12)(-6) + 1 = -143 \\ f(4) &= (4)^3 - (12)(4) + 1 = 17 \\ f(-2) &= (-2)^3 - (12)(-2) + 1 = 17 \\ f(2) &= (2)^3 - (12)(2) + 1 = -15 \end{aligned}$$

The minimum value of the function $f(x)$ is -143, occurring at the endpoint $x = -6$.

The maximum value of the function $f(x)$ is 17, occurring at both the critical point, $x = -2$ (local maximum), and at the endpoint, $x = 4$.

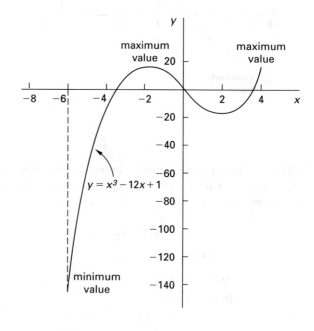

(b) The maximum and minimum values occur either at local extrema or at the endpoints. The critical points are located where the first derivative is zero.

$$\begin{aligned} f(x) &= x + 1 \\ f'(x) &= 1 \end{aligned}$$

$f'(x)$ is never equal to zero. There are no critical points.

The extreme values of the function $f(x)$ must occur at the endpoints, -5 and 5, of the interval.

$$\begin{aligned} f(-5) &= -5 + 1 = -4 \\ f(5) &= 5 + 1 = 6 \end{aligned}$$

The minimum value of the function $f(x)$ is -4, occurring at the endpoint $x = -5$.

The maximum value of the function $f(x)$ is 6, occurring at the endpoint $x = 5$.

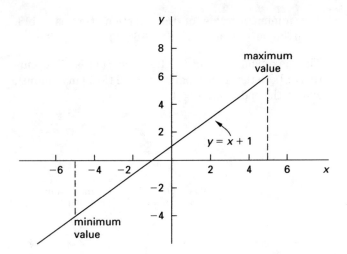

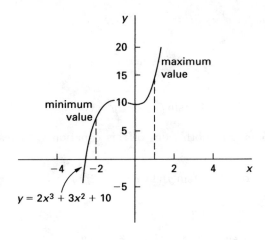

7. The maximum and minimum values occur either at local extrema or at the endpoints. The critical points are located where the first derivative is zero.

$$f(x) = 2x^3 + 3x^2 + 10$$
$$f'(x) = 6x^2 + 6x$$
$$6x^2 + 6x = 0$$
$$6x(x + 1) = 0$$
$$x = -1, 0$$

Test for a local maximum, local minimum, or inflection point.

$$f''(x) = 12x + 6$$
$$f''(-1) = (12)(-1) + 6 = -6 < 0 \quad \text{[local maximum]}$$
$$f''(0) = (12)(0) + 6 = 6 > 0 \quad \text{[local minimum]}$$

Check the value of the function $f(x)$ at the endpoints, -2 and 1, of the interval, and compare with the value of $f(x)$ at the critical points.

$$f(-2) = (2)(-2)^3 + (3)(-2)^2 + 10 = 6$$
$$f(1) = (2)(1)^3 + (3)(1)^2 + 10 = 15$$
$$f(-1) = (2)(-1)^3 + (3)(-1)^2 + 10 = 11$$
$$f(0) = (2)(0)^3 + (3)(0) + 10 = 10$$

The minimum value of the function $f(x)$ is 6, occurring at the endpoint $x = -2$.

The maximum value of the function $f(x)$ is 15, occurring at the endpoint $x = 1$.

8. (a) The critical points are located where the first derivative is zero.

$$f(x) = x - \sqrt{2}\cos x$$
$$f'(x) = 1 + \sqrt{2}\sin x$$
$$1 + \sqrt{2}\sin x = 0$$
$$\sin x = -\frac{1}{\sqrt{2}}$$
$$x = \arcsin\left(-\frac{1}{\sqrt{2}}\right)$$

On the interval $[-(\pi/2), 0]$, this is satisfied by $x = -(\pi/4)$.

Test for a local maximum, local minimum, or inflection point.

$$f''(x) = \sqrt{2}\cos x$$
$$f''\left(-\frac{\pi}{4}\right) = \sqrt{2}\cos\left(-\frac{\pi}{4}\right)$$
$$= \sqrt{2}\left(\frac{1}{\sqrt{2}}\right) = 1 > 0 \quad \text{[local minimum]}$$

It follows that there is a local minimum at $x = -(\pi/4)$.

(b) The (absolute) maximum and (absolute) minimum values occur either at the local minimum found in part (a) or at the endpoints. Check the value of the function $f(x)$ at the endpoints, $-(\pi/2)$ and 0, of the interval, and compare with the value of $f(x)$ at the critical point, $x = -(\pi/4)$.

$$f\left(-\frac{\pi}{2}\right) = \left(-\frac{\pi}{2}\right) - \sqrt{2}\cos\left(-\frac{\pi}{2}\right)$$
$$= -\frac{\pi}{2}$$
$$\sim -1.57$$

$$f(0) = 0 - \sqrt{2}\cos 0$$
$$= -\sqrt{2}$$
$$\sim -1.41$$

$$f\left(-\frac{\pi}{4}\right) = \left(-\frac{\pi}{4}\right) - \sqrt{2}\cos\left(-\frac{\pi}{4}\right)$$
$$= \left(-\frac{\pi}{4}\right) - \sqrt{2}\left(\frac{1}{\sqrt{2}}\right)$$
$$= -\frac{\pi}{4} - 1$$
$$\sim -1.79$$

The minimum value of the function $f(x)$ is $-(\pi/4) - 1 \sim -1.79$, occurring at the critical point $x = -(\pi/4)$.

The maximum value of the function $f(x)$ is $-\sqrt{2} \sim -1.41$, occurring at the endpoint $x = 0$.

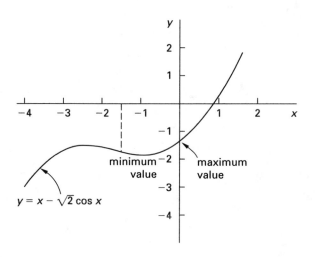

$y = x - \sqrt{2}\cos x$

There is a possible point of inflection at $x = 0$. Check to see if $f''(x)$ changes sign at $x = 0$ (as in Ex. 1.25).

sample x:	-0.5	0	0.5
sign of $f''(x)$:	$+$	0	$-$

The second derivative, $f''(x)$, does change sign at $x = 0$, so $x = 0$ is indeed a point of inflection.

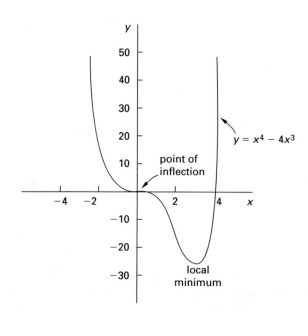

$y = x^4 - 4x^3$

point of inflection

local minimum

9. Critical points occur where the first derivative is zero.

$$f(x) = x^4 - 4x^3$$
$$f'(x) = 4x^3 - 12x^2$$
$$4x^3 - 12x^2 = 0$$
$$4x^2(x - 3) = 0$$
$$x = 0, 3$$

Test for a local maximum, local minimum, or inflection point.

$$f''(x) = 12x^2 - 24x$$
$$f''(0) = 0 - 0 = 0$$
$$f''(3) = (12)(3)^2 - (24)(3) = 36 > 0 \quad \text{[local minimum]}$$

10. To find $\partial f/\partial x$, treat y as constant. (All terms that do not contain x have zero derivatives.) To find $\partial f/\partial y$, treat x as constant. (All terms that do not contain y have zero derivatives.)

(a)

$$f(x, y) = x^3y^2 + yx^4 + \sin y + \cos^2 y + \sin^3 x$$

$$\frac{\partial f}{\partial x} = 3x^2y^2 + 4yx^3 + 0 + 0 + 3\sin^2 x \cos x$$
$$= 3x^2y^2 + 4yx^3 + 3\sin^2 x \cos x$$

$$\frac{\partial f}{\partial y} = 2x^3y + x^4 + \cos y + 2\cos y(-\sin y) + 0$$
$$= 2x^3y + x^4 + \cos y - 2\cos y \sin y$$

(b)

$$f(x,y) = \ln(xy) + \frac{x}{y} = \ln x + \ln y + \frac{x}{y} \quad \text{[see App. 3]}$$

$$\frac{\partial f}{\partial x} = \frac{1}{x} + 0 + \frac{1}{y}$$

$$= \frac{1}{x} + \frac{1}{y}$$

$$\frac{\partial f}{\partial y} = 0 + \frac{1}{y} + x(-y^{-2})$$

$$= \frac{1}{y} - \frac{x}{y^2}$$

(c) Use the quotient rule (Eq. 1.6) for partial differentiation.

$$f(x,y) = \frac{x^3 - y^2}{x + y}$$

$$\frac{\partial f}{\partial x} = \frac{(x+y)(3x^2) - (x^3 - y^2)(1)}{(x+y)^2}$$

$$= \frac{3x^2(x+y) - (x^3 - y^2)}{(x+y)^2}$$

$$\frac{\partial f}{\partial y} = \frac{(x+y)(-2y) - (x^3 - y^2)(1)}{(x+y)^2}$$

$$= \frac{-2y(x+y) - (x^3 - y^2)}{(x+y)^2}$$

11. To find $\partial f/\partial x$, treat y and z as constant. (All terms that do not contain x have zero derivatives.) To find $\partial f/\partial y$, treat x and z as constant. (All terms that do not contain y have zero derivatives.) To find $\partial f/\partial z$, treat y and x as constant. (All terms that do not contain z have zero derivatives.)

(a)

$$f(x,y,z) = z \ln(x+y+z)$$

$$\frac{\partial f}{\partial x} = z\frac{\partial}{\partial x}[\ln(x+y+z)]$$

As in Ex. 1.32, let $u = x+y+z$ and use the chain rule for partial differentiation.

$$\frac{\partial}{\partial x}\left(\ln(x+y+z)\right) = \frac{d}{du}(\ln u)\frac{\partial u}{\partial x}$$

$$= \left(\frac{1}{u}\right)(1) = \frac{1}{x+y+z}$$

Finally,

$$\frac{\partial f}{\partial x} = z\frac{\partial}{\partial x}\left(\ln(x+y+z)\right)$$

$$= \frac{z}{x+y+z}$$

Similarly,

$$\frac{\partial f}{\partial y} = z\frac{\partial}{\partial y}\left(\ln(x+y+z)\right)$$

$$= z\frac{d}{du}(\ln u)\frac{\partial u}{\partial y}$$

$$= z\left(\frac{1}{u}\right)(1)$$

$$= \frac{z}{x+y+z}$$

To find $\partial f/\partial z$, use the product rule (Eq. 1.5) for partial differentiation.

$$f(x,y,z) = z\ln(x+y+z)$$

$$\frac{\partial f}{\partial z} = z\frac{\partial}{\partial z}\left(\ln(x+y+z)\right) + [\ln(x+y+z)](1)$$

$$= z\frac{d}{du}(\ln u)\frac{\partial u}{\partial z} + \ln(x+y+z)$$

$$= z\left(\frac{1}{u}\right)(1) + \ln(x+y+z)$$

$$= \frac{z}{x+y+z} + \ln(x+y+z)$$

(b)

$$f(x,y,z) = 3y^2 + 6xz + \frac{x}{\ln z} + \sin(x^2 + z)$$

$$\frac{\partial f}{\partial x} = 0 + 6z + \left(\frac{1}{\ln z}\right)(1) + [\cos(x^2 + z)](2x)$$

$$= 6z + \frac{1}{\ln z} + 2x\cos(x^2 + z)$$

$$\frac{\partial f}{\partial y} = 6y + 0 + 0 + [\cos(x^2 + z)](0)$$

$$= 6y$$

$$\frac{\partial f}{\partial z} = 0 + 6x + x\left[-\frac{1}{z(\ln z)^2}\right] + [\cos(x^2 + z)](1)$$

$$= 6x - \frac{x}{z(\ln z)^2} + \cos(x^2 + z)$$

12.

$$G(x,y) = \ln(x+y) - \sin(x-y) + e^{xy}$$

$$G_x = \frac{1}{x+y} - \cos(x-y) + ye^{xy}$$

$$G_y = \frac{1}{x+y} - [\cos(x-y)](-1) + xe^{xy}$$

$$= \frac{1}{x+y} + \cos(x-y) + xe^{xy}$$

$$G_{xx} = \frac{\partial}{\partial x}(G_x)$$

$$= \frac{\partial}{\partial x}\left(\frac{1}{x+y} - \cos(x-y) + ye^{xy}\right)$$

$$= -(x+y)^{-2} - [-\sin(x-y)] + y(ye^{xy})$$

$$= -\frac{1}{(x+y)^2} + \sin(x-y) + y^2 e^{xy}$$

$$G_{xy} = \frac{\partial}{\partial y}(G_x)$$

$$= \frac{\partial}{\partial y}\left(\frac{1}{x+y} - \cos(x-y) + ye^{xy}\right)$$

$$= -(x+y)^{-2} - [(-\sin(x-y))(-1)]$$
$$\quad + [y(xe^{xy}) + e^{xy}(1)]$$

$$= -\frac{1}{(x+y)^2} - \sin(x-y) + e^{xy}(1+xy)$$

Similarly,

$$G_{yy} = \frac{\partial}{\partial y}(G_y)$$

$$= \frac{\partial}{\partial y}\left(\frac{1}{x+y} + \cos(x-y) + xe^{xy}\right)$$

$$= -(x+y)^{-2} + [(-\sin(x-y))(-1)] + x(xe^{xy})$$

$$= -\frac{1}{(x+y)^2} + \sin(x-y) + x^2 e^{xy}$$

$$G_{yx} = \frac{\partial}{\partial x}(G_y)$$

$$= \frac{\partial}{\partial x}\left(\frac{1}{x+y} + \cos(x-y) + xe^{xy}\right)$$

$$= -(x+y)^{-2} + [(-\sin(x-y))(1)]$$
$$\quad + [x(ye^{xy}) + e^{xy}(1)]$$

$$= -\frac{1}{(x+y)^2} - \sin(x-y) + e^{xy}(xy+1)$$

$$= G_{xy} \quad \text{[as required]}$$

13.

$$K = \left|\frac{y''}{\left[1+(y')^2\right]^{\frac{3}{2}}}\right|$$

Let $y = \ln x$. Then,

$$y' = \frac{1}{x}$$

$$y'' = -\frac{1}{x^2}$$

$$K = \left|\frac{y''}{\left[1+(y')^2\right]^{\frac{3}{2}}}\right|$$

$$= \left|\frac{-\dfrac{1}{x^2}}{\left[1+\left(\dfrac{1}{x}\right)^2\right]^{\frac{3}{2}}}\right|$$

At the point $(2, \ln 2)$, $x = 2$. Consequently,

$$K = \left|\frac{-\dfrac{1}{4}}{\left[1+\left(\dfrac{1}{2}\right)^2\right]^{\frac{3}{2}}}\right|$$

$$= \left|\left(-\frac{1}{4}\right)\left(\frac{4}{5}\right)^{\frac{3}{2}}\right|$$

$$\sim 0.179$$

14.

$$K = \left|\frac{y''}{\left[1+(y')^2\right]^{\frac{3}{2}}}\right|$$

Let $y = x^2$. Then,

$$y' = 2x$$

$$y'' = 2$$

$$K = \left|\frac{2}{\left[1+(2x)^2\right]^{\frac{3}{2}}}\right|$$

$$= \frac{2}{\left[1+4x^2\right]^{\frac{3}{2}}}$$

To find the maximum value of K, find the critical points.

$$\frac{dK}{dx} = (2)\left(-\frac{3}{2}\right)(1+4x^2)^{-\frac{5}{2}}(8x)$$

$$= -(3)(1+4x^2)^{-\frac{5}{2}}(8x)$$

Set dK/dx equal to zero.

$$-(3)(1+4x^2)^{-\frac{5}{2}}(8x) = 0$$

This equation is satisfied by $x = 0$. To show that this corresponds to a maximum, find the second derivative.

$$\frac{d^2 K}{dx^2} = -(24)\left((1+4x^2)^{-\frac{5}{2}}(1)\right.$$
$$\left. + (x)\left[-\frac{5}{2}(1+4x^2)^{-\frac{7}{2}}(8x)\right]\right)$$
$$= -24 \quad [\text{when } x = 0]$$

It follows that $x = 0$ corresponds to a local maximum of K. In fact, since $K = 2/(1+4x^2)^{\frac{3}{2}}$, the denominator achieves its smallest value when $x = 0$. Consequently, K attains its (global) maximum value at $x = 0$. It follows that the parabola $y = x^2$ has maximum curvature at the origin.

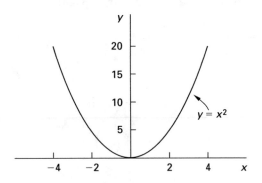

For Probs. 15 through 20, before applying l'Hôpital's rule, check that the limit is indeed indeterminate of the form $\frac{0}{0}$ or $\frac{\infty}{\infty}$.

15. Let $x = -1$. The limit is indeterminate of the form $\frac{0}{0}$. Apply l'Hôpital's rule.

$$\lim_{x \to -1} \frac{x^3 + 4x^2 + 5x + 2}{x^2 + 2x + 1} = \lim_{x \to -1} \frac{3x^2 + 8x + 5}{2x + 2}$$

The limit continues to be indeterminate of the form $\frac{0}{0}$. Apply l'Hôpital's rule once more.

$$\lim_{x \to -1} \frac{3x^2 + 8x + 5}{2x + 2} = \lim_{x \to -1} \frac{6x + 8}{2} = \frac{2}{2} = 1$$

16. Let $x = 0$. The limit becomes

$$\frac{\sqrt{1+0} - 3}{\sqrt{0} - 1} = \frac{1 - 3}{-1} = 2$$

17. Let $x = \infty$. The limit is indeterminate of the form $\frac{\infty}{\infty}$. Apply l'Hôpital's rule.

$$\lim_{x \to \infty} \frac{3x^2}{\pi e^x} = \lim_{x \to \infty} \frac{6x}{\pi e^x}$$

The limit continues to be indeterminate of the form $\frac{\infty}{\infty}$. Apply l'Hôpital's rule once more.

$$\lim_{x \to \infty} \frac{6x}{\pi e^x} = \lim_{x \to \infty} \frac{6}{\pi e^x} = \frac{6}{\infty} = 0$$

18. Let $x = \infty$.

$$\frac{\sin\left(\dfrac{2}{x}\right)}{\sin\left(\dfrac{3}{x}\right)} = \frac{\sin 0}{\sin 0} = \frac{0}{0}$$

The limit is indeterminate of the form $\frac{0}{0}$. Apply l'Hôpital's rule.

$$\lim_{x \to \infty} \frac{\sin\left(\dfrac{2}{x}\right)}{\sin\left(\dfrac{3}{x}\right)} = \lim_{x \to \infty} \frac{\left(\dfrac{-2}{x^2}\right)\cos\left(\dfrac{2}{x}\right)}{\left(\dfrac{-3}{x^2}\right)\cos\left(\dfrac{3}{x}\right)}$$
$$= \left(\frac{2}{3}\right)\lim_{x \to \infty} \frac{\cos\left(\dfrac{2}{x}\right)}{\cos\left(\dfrac{3}{x}\right)}$$
$$= \left(\frac{2}{3}\right)\left(\frac{\cos 0}{\cos 0}\right)$$
$$= \frac{2}{3}$$

19. Let $\theta = 0$.

$$\frac{\sin^2 \theta}{\theta} = \frac{\sin 0}{0} = \frac{0}{0}$$

The limit is indeterminate of the form $\frac{0}{0}$. Apply l'Hôpital's rule.

$$\lim_{\theta \to 0} \frac{\sin^2 \theta}{\theta} = \lim_{\theta \to 0} \frac{2 \sin \theta \cos \theta}{1}$$
$$= \lim_{\theta \to 0} 2 \sin \theta \cos \theta$$
$$= (2)(0)(1) = 0$$

20. Let $x = \infty$. The limit is indeterminate of the form $\frac{\infty}{\infty}$. Apply l'Hôpital's rule.

$$\lim_{x \to \infty} \frac{x^{10} + 1}{x - 3} = \lim_{x \to \infty} \frac{10x^9}{1}$$
$$= \lim_{x \to \infty} 10x^9 = \infty$$

The limit does not exist.

Chapter Two

1.

$$A = \text{area} = \int_{-1}^{2} (x^2 + 1)\, dx$$

$$= \left[\frac{x^3}{3} + x \right]_{-1}^{2}$$

$$= \left[\frac{(2)^3}{3} + 2 \right] - \left[\frac{(-1)^3}{3} + (-1) \right]$$

$$= 6$$

2. In what follows, C is an arbitrary constant of integration.

(a)

$$\int u(\sqrt{u} + u^3)\, du = \int (u^{\frac{3}{2}} + u^4)\, du$$

$$= \frac{u^{\frac{5}{2}}}{\frac{5}{2}} + \frac{u^5}{5} + C$$

$$= \frac{2}{5} u^{\frac{5}{2}} + \frac{u^5}{5} + C$$

(b)

$$\int (x+1)^2 dx = \int (x^2 + 2x + 1)\, dx$$

$$= \frac{x^3}{3} + (2)\left(\frac{x^2}{2} \right) + x + C$$

$$= \frac{x^3}{3} + x^2 + x + C$$

(c)

$$\int \left(\frac{t^5 - t}{\sqrt{t}} \right) dt = \int (t^{\frac{9}{2}} - t^{\frac{1}{2}})\, dt$$

$$= \frac{t^{\frac{11}{2}}}{\frac{11}{2}} - \frac{t^{\frac{3}{2}}}{\frac{3}{2}} + C$$

$$= \frac{2}{11} t^{\frac{11}{2}} - \frac{2}{3} t^{\frac{3}{2}} + C$$

3. (a)

$$\int_{-\frac{\pi}{2}}^{\frac{\pi}{2}} \cos x\, dx = \left[\sin x \right]_{-\frac{\pi}{2}}^{\frac{\pi}{2}}$$

$$= \sin \left(\frac{\pi}{2} \right) - \sin \left(-\frac{\pi}{2} \right)$$

$$= 1 - (-1) = 2$$

(b)

$$\int_{1}^{3} x^{-3} dx = \left[\frac{x^{-2}}{-2} \right]_{1}^{3}$$

$$= -\left(\frac{1}{2} \right) \left[\frac{1}{(3)^2} - \frac{1}{(1)^2} \right]$$

$$= \frac{4}{9}$$

(c)

$$\int_{-1}^{0} (t-1)(2t+2)\, dt = \int_{-1}^{0} (2t^2 - 2)\, dt$$

$$= \left[(2)\left(\frac{t^3}{3} \right) - 2t \right]_{-1}^{0}$$

$$= \left[(2)\left(\frac{(0)^3}{3} \right) - (2)(0) \right]$$

$$\quad - \left[(2)\left(\frac{(-1)^3}{3} \right) - (2)(-1) \right]$$

$$= -\frac{4}{3}$$

For Probs. 4 through 9, C is an arbitrary constant of integration.

4. (a) Use Eq. 2.8 with $f(x) = \ln x$, $h(x) = 1/x^2$, and $H(x) = -(1/x)$.

$$\int \left(\frac{\ln x}{x^2} \right) dx = (\ln x)\left(-\frac{1}{x} \right) - \int \left(-\frac{1}{x} \right)\left(\frac{1}{x} \right) dx$$

$$= -\left(\frac{1}{x} \right) \ln x + \int x^{-2}\, dx$$

$$= -\left(\frac{1}{x} \right) \ln x + \frac{x^{-1}}{-1} + C$$

$$= -\left(\frac{1}{x} \right)(\ln x + 1) + C$$

(b) Use Eq. 2.8 with $f(t) = \ln t$, $h(t) = t^{\frac{1}{2}}$, and $H(t) = (2/3)t^{\frac{3}{2}}$.

$$\int t^{\frac{1}{2}} \ln t\, dt = \ln t \left[\left(\frac{2}{3} \right) t^{\frac{3}{2}} \right] - \int \left[\left(\frac{2}{3} \right) t^{\frac{3}{2}} \right]\left(\frac{1}{t} \right) dt$$

$$= \left(\frac{2}{3} \right) t^{\frac{3}{2}} \ln t - \frac{2}{3} \int t^{\frac{1}{2}}\, dt$$

$$= \left(\frac{2}{3} \right) t^{\frac{3}{2}} \ln t - \left(\frac{2}{3} \right)\left(\frac{t^{\frac{3}{2}}}{\frac{3}{2}} \right) + C$$

$$= \left(\frac{2}{3} \right) t^{\frac{3}{2}} \ln t - \left(\frac{4}{9} \right) t^{\frac{3}{2}} + C$$

$$= \left(\frac{2}{3} \right) t^{\frac{3}{2}} \left(\ln t - \frac{2}{3} \right) + C$$

(c) Use Eq. 2.9 with $f(x) = x$, $h(x) = e^{2x}$, and $H(x) = (1/2)e^{2x}$.

$$\int_0^1 xe^{2x}\,dx = \left[x\frac{1}{2}e^{2x}\right]_0^1 - \int_0^1 \left(\frac{1}{2}e^{2x}\right)(1)\,dx$$

$$= \left(\frac{1}{2}\right)[1e^2 - 0e^0] - \frac{1}{2}\int_0^1 e^{2x}\,dx$$

$$= \left(\frac{1}{2}\right)\left(e^2 - \left(\frac{1}{2}\right)[e^{2x}]_0^1\right)$$

$$= \left(\frac{1}{2}\right)\left[e^2 - \left(\frac{1}{2}\right)(e^2 - e^0)\right]$$

$$= \left(\frac{1}{2}\right)\left(\frac{1}{2}e^2 + \frac{1}{2}\right)$$

$$= \left(\frac{1}{4}\right)(e^2 + 1)$$

(d) Use Eq. 2.8 with $f(t) = t^2$, $h(t) = \sin t$, and $H(t) = -\cos t$.

$$\int t^2 \sin t\,dt = t^2(-\cos t) - \int (-\cos t)(2t)\,dt$$

$$= -t^2 \cos t + 2\int t\cos t\,dt$$

Apply Eq. 2.8 (with $f(t) = t$ and $h(t) = \cos t$) to the resulting integral.

$$\int t^2 \sin t\,dt = -t^2 \cos t + 2\int t\cos t\,dt$$

$$= -t^2 \cos t + (2)\left[t\sin t - \int (\sin t)(1)\,dt\right]$$

$$= -t^2 \cos t + (2)[t\sin t - (-\cos t)] + C$$

$$= -t^2 \cos t + 2t\sin t + 2\cos t + C$$

5. (a)

$$u = \cos x$$

$$du = -\sin x\,dx$$

$$\int \cos^2 x \sin x\,dx = -\int u^2\,du$$

$$= -\frac{u^3}{3} + C$$

$$= -\frac{\cos^3 x}{3} + C$$

(b)

$$u = x^2 + 1$$

$$du = 2x\,dx$$

$$\int x(x^2 + 1)^{25}\,dx = \frac{1}{2}\int u^{25}\,du$$

$$= \left(\frac{1}{2}\right)\left(\frac{u^{26}}{26}\right) + C$$

$$= \frac{(x^2 + 1)^{26}}{52} + C$$

6. In each case, the integral is of the form of the integral in Eq. 2.10. The substitution is $u = g(x)$ where $g(x)$ is found, as in Exs. 2.19 through 2.24, by identifying a function $g(x)$ whose differential $g'(x)\,dx$ also appears in the integral.

(a) Let $u = x^6 + 10x$.

$$du = (6x^5 + 10)\,dx$$

$$\int \left(\frac{6x^5 + 10}{\sqrt{x^6 + 10x}}\right)dx = \int \frac{du}{\sqrt{u}}$$

$$= \int u^{-\frac{1}{2}}\,du$$

$$= \frac{u^{\frac{1}{2}}}{\frac{1}{2}} + C$$

$$= (2)(x^6 + 10x)^{\frac{1}{2}} + C$$

(b) Let $u = 2x^4 + 10$.

$$du = 8x^3\,dx$$

$$\int 4x^3 \cos(2x^4 + 10)\,dx = \frac{1}{2}\int \cos u\,du$$

$$= \frac{1}{2}\sin u + C$$

$$= \frac{1}{2}\sin(2x^4 + 10) + C$$

(c) Let $u = x^2 + 9$.

$$du = 2x\,dx$$

Since this is a definite integral, change the "x-limits" to "u-limits."

$$x = 0$$
$$u = (0)^2 + 9 = 9$$
$$x = 4$$
$$u = (4)^2 + 9 = 25$$

The integral becomes

$$\frac{1}{2}\int_9^{25} u^{\frac{1}{2}}\,du = \left(\frac{1}{2}\right)\left[\frac{u^{\frac{3}{2}}}{\frac{3}{2}}\right]_9^{25}$$

$$= \left(\frac{1}{3}\right)\left[(25)^{\frac{3}{2}} - (9)^{\frac{3}{2}}\right]$$

$$= \left(\frac{1}{3}\right)(125 - 27)$$

$$= \frac{98}{3}$$

(d) Let $u = x^2 + 2x + 1$.

$$du = (2x + 2)\,dx = (2)(x + 1)\,dx$$

Since this is a definite integral, change the "x-limits" to "u-limits."

$$x = -1$$
$$u = (-1)^2 + (2)(-1) + 1 = 0$$
$$x = 1$$
$$u = (1)^2 + (2)(1) + 1 = 4$$

The integral becomes

$$\frac{1}{2}\int_0^4 u^{-\frac{1}{2}}\,du = \left(\frac{1}{2}\right)\left[\frac{u^{\frac{1}{2}}}{\frac{1}{2}}\right]_0^4$$
$$= \left[u^{\frac{1}{2}}\right]_0^4$$
$$= (4)^{\frac{1}{2}} - (0)^{\frac{1}{2}}$$
$$= 2$$

Alternatively, note that

$$\frac{x + 1}{\sqrt{x^2 + 2x + 1}} = \frac{x + 1}{\sqrt{(x+1)^2}} = \frac{x + 1}{x + 1} = 1$$
$$\int_{-1}^{1}\left(\frac{x + 1}{\sqrt{x^2 + 2x + 1}}\right)dx = \int_{-1}^{1} 1\,dx = \left[x\right]_{-1}^{1} = 2$$

7. According to Eq. 2.12, let $x = 2\sin\theta$. Then $dx = 2\cos\theta\,d\theta$.

$$\int\left(\frac{\sqrt{4 - x^2}}{x}\right)dx = \int\left(\frac{\sqrt{4 - 4\sin^2\theta}}{2\sin\theta}\right)2\cos\theta\,d\theta$$
$$= \int\left(\frac{2\sqrt{1 - \sin^2\theta}}{2\sin\theta}\right)2\cos\theta\,d\theta$$
$$= \int\left(\frac{2\sqrt{\cos^2\theta}}{2\sin\theta}\right)2\cos\theta\,d\theta$$
$$= \int\left(\frac{4\cos^2\theta}{2\sin\theta}\right)d\theta$$
$$= 2\int\left(\frac{1 - \sin^2\theta}{\sin\theta}\right)d\theta$$
$$= 2\int(\csc\theta - \sin\theta)\,d\theta$$

From Table 2.2,

$$2\int(\csc\theta - \sin\theta)d\theta = (2)(\ln|\csc\theta - \cot\theta| + \cos\theta) + C$$

It follows that

$$\int\left(\frac{\sqrt{4 - x^2}}{x}\right)dx = (2)(\ln|\csc\theta - \cot\theta| + \cos\theta) + C$$

Return to the original variable x by noting that from the following figure,

$$x = 2\sin\theta$$
$$\sin\theta = \frac{x}{2}$$
$$\cos\theta = \frac{\sqrt{4 - x^2}}{2}$$
$$\cot\theta = \frac{1}{\tan\theta} = \frac{\sqrt{4 - x^2}}{x}$$
$$\csc\theta = \frac{1}{\sin\theta} = \frac{2}{x}$$

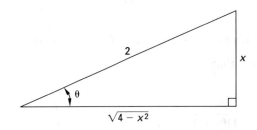

The integral becomes

$$\int\left(\frac{\sqrt{4 - x^2}}{x}\right)dx = (2)(\ln|\csc\theta - \cot\theta| + \cos\theta) + C$$
$$= (2)\left(\ln\left|\frac{2}{x} - \frac{\sqrt{4 - x^2}}{x}\right|\right.$$
$$\left. + \frac{\sqrt{4 - x^2}}{2}\right) + C$$

8. (a)

$$\frac{1}{x^3 - x} = \frac{1}{x(x^2 - 1)} = \frac{1}{x(x - 1)(x + 1)}$$

Decompose this expression into partial fractions using Eq. 2.15.

$$\frac{1}{x(x - 1)(x + 1)} = \frac{A}{x} + \frac{B}{x - 1} + \frac{C}{x + 1}$$

Here, A, B, and C are constants to be determined. Clear the fractions.

$$1 = A(x - 1)(x + 1) + Bx(x + 1) + Cx(x - 1)$$
$$= (A + B + C)x^2 + (B - C)x - A$$

Equate the coefficients of terms with different powers of x.

$$A + B + C = 0$$
$$B - C = 0$$
$$-A = 1$$

Solving this system leads to $A = -1$ and $B = C = 1/2$.

Alternatively, substitute, in turn, $x = -1$, $x = 0$, and $x = 1$ in the equation.

$$1 = A(x-1)(x+1) + Bx(x+1) + Cx(x-1)$$

The following set of equations is obtained.

$$1 = 2C$$
$$1 = -A$$
$$1 = 2B$$

This leads to $A = -1$ and $B = C = 1/2$, as above.

It follows that

$$\frac{1}{x(x-1)(x+1)} = \frac{-1}{x} + \frac{\frac{1}{2}}{x-1} + \frac{\frac{1}{2}}{x+1}$$

$$\int \frac{dx}{x^3 - x} = \int \frac{dx}{x(x-1)(x+1)}$$

$$= \int \left(\frac{-1}{x} + \frac{\frac{1}{2}}{x-1} + \frac{\frac{1}{2}}{x+1} \right) dx$$

$$= -\ln|x| + \frac{1}{2}\ln|x-1|$$
$$\quad + \frac{1}{2}\ln|x+1| + C_1$$

$$= \left(\frac{1}{2} \right)(\ln|x-1||x+1|)$$
$$\quad - \ln|x| + C_1$$

$$= \ln \left| \frac{(x^2-1)^{\frac{1}{2}}}{x} \right| + C_1$$

C_1 is an arbitrary constant of integration.

(b)
$$\frac{x+1}{x^3 - x^2} = \frac{x+1}{x^2(x-1)}$$

Decompose this expression into partial fractions using Eqs. 2.15 and 2.16.

$$\frac{x+1}{x^2(x-1)} = \frac{A}{x} + \frac{B}{x^2} + \frac{C}{x-1}$$

Here, A, B, and C are constants to be determined. Clear the fractions.

$$x + 1 = Ax(x-1) + B(x-1) + Cx^2$$
$$= (A+C)x^2 + (B-A)x - B$$

Equate the coefficients of terms with different powers of x.

$$A + C = 0$$
$$B - A = 1$$
$$-B = 1$$

Solving this system leads to $A = -2$, $B = -1$, and $C = 2$.

Alternatively, substitute, in turn, $x = 0$ and $x = 1$ into the equation.

$$x + 1 = Ax(x-1) + B(x-1) + Cx^2$$

The following equations are obtained.

$$1 = -B$$
$$2 = C$$

The constant A is obtained from the following equation (derived above) in which the values $C = 2$ and $B = -1$ have been used.

$$A + C = 0$$
$$A + 2 = 0$$
$$A = -2$$

This again leads to the conclusion that $A = -2$, $B = -1$, and $C = 2$.

It follows that

$$\frac{x+1}{x^2(x-1)} = \frac{-2}{x} - \frac{1}{x^2} + \frac{2}{x-1}$$

$$\int \left(\frac{x+1}{x^2(x-1)} \right) dx = \int \left(\frac{-2}{x} - \frac{1}{x^2} + \frac{2}{x-1} \right) dx$$

$$= \int \left(\frac{-2}{x} - \frac{1}{x^2} + \frac{2}{x-1} \right) dx$$

$$= -2\ln|x| + \frac{1}{x} + 2\ln|x-1| + C_1$$

$$= (2)(\ln|x-1| - \ln|x|) + \frac{1}{x} + C_1$$

$$= 2\ln \left| \frac{x-1}{x} \right| + \frac{1}{x} + C_1$$

C_1 is an arbitrary constant of integration.

(c) Decompose the integrand into partial fractions using Eqs. 2.15 and 2.17.

$$\frac{x}{(x+1)(x^2+1)} = \frac{A}{x+1} + \frac{Bx+C}{x^2+1}$$

Here, A, B, and C are constants to be determined. Clear the fractions.

$$x = A(x^2 + 1) + (Bx + C)(x + 1)$$
$$= (A + B)x^2 + (B + C)x + A + C$$

Equate the coefficients of terms with different powers of x.

$$A + B = 0$$
$$B + C = 1$$
$$A + C = 0$$

Solving this system (using Cramer's rule or elimination) leads to $A = -1/2$, $B = 1/2$, and $C = 1/2$.

Alternatively, let $x = -1$ in the equation.

$$x = A(x^2 + 1) + (Bx + C)(x + 1)$$

This leads to $-1 = 2A$ or $A = -(1/2)$. The constants B and C are obtained from the following equations (derived above) using the value $A = -(1/2)$.

$$A + B = 0$$
$$A + C = 0$$
$$-\frac{1}{2} + B = 0$$
$$-\frac{1}{2} + C = 0$$

This again leads to the conclusion that $A = -(1/2)$ and $B = C = 1/2$.

It follows that

$$\frac{x}{(x + 1)(x^2 + 1)} = \left(\frac{1}{2}\right)\left(\frac{-1}{x + 1} + \frac{x + 1}{x^2 + 1}\right)$$

$$\int \left(\frac{x}{(x + 1)(x^2 + 1)}\right) dx$$
$$= \frac{1}{2} \int \left(\frac{-1}{x + 1} + \frac{x}{x^2 + 1} + \frac{1}{x^2 + 1}\right) dx$$
$$= \left(\frac{1}{2}\right)\left(-\ln|x + 1| + \frac{1}{2}\ln|x^2 + 1| + \arctan x\right) + C_1$$

The substitution $u = x^2 + 1$ has been used to determine the integral $\int [(x)/(x^2 + 1)]dx$ (see Prob. 9). The other two integrals come from Tables 2.2 and 2.3. Finally,

$$\int \left(\frac{x}{(x + 1)(x^2 + 1)}\right) dx$$
$$= \left(\frac{1}{2}\right)\left(\ln\left|\frac{(x^2 + 1)^{\frac{1}{2}}}{x + 1}\right| + \arctan x\right) + C_1$$

C_1 is an arbitrary constant of integration.

9. Use integration by parts (Eq. 2.8) with $f(x) = \arctan x$ and $h(x) = 1$.

$$\int \arctan x \, dx = \int (1) \arctan x \, dx$$
$$= (\arctan x)x - \int \left(\frac{1}{1 + x^2}\right)(x) \, dx$$
$$= x \arctan x - \int \left(\frac{x}{1 + x^2}\right) dx$$

Let $u = 1 + x^2$ in the last integral. Then $du = 2x \, dx$.

$$\int \frac{x}{1 + x^2} \, dx = \frac{1}{2} \int \frac{du}{u}$$
$$= \frac{1}{2} \ln|u| + C_1$$
$$= \frac{1}{2} \ln|1 + x^2| + C_1$$

Finally,

$$\int \arctan x \, dx = x \arctan x - \frac{1}{2} \ln|1 + x^2| + C_1$$

C_1 is an arbitrary constant of integration.

Chapter Three

1. (a) Proceed as in Exs. 3.1 and 3.5. From Eq. 3.4, with $a = 0$, $b = \sqrt{2}$, and $f(x) = x^2 + 1$ (see the figure in part (a) of problem),

$$A = \int_0^{\sqrt{2}} (x^2 + 1) \, dx$$
$$= \left[\frac{x^3}{3} + x\right]_0^{\sqrt{2}} = \frac{2\sqrt{2}}{3} + \sqrt{2} - 0$$
$$= \left(\frac{5}{3}\right)\sqrt{2}$$

The centroid is located at (x_c, y_c).

$$x_c = \frac{1}{A} \int_0^{\sqrt{2}} x(x^2 + 1) \, dx$$
$$= \left(\frac{1}{A}\right)\left[\frac{x^4}{4} + \frac{x^2}{2}\right]_0^{\sqrt{2}}$$
$$= \left(\frac{1}{A}\right)(1 + 1 - 0)$$
$$= \frac{2}{A}$$
$$= \frac{6}{5\sqrt{2}}$$
$$= \frac{3}{5}\sqrt{2}$$

$$y_c = \frac{1}{2A} \int_0^{\sqrt{2}} (x^2 + 1)^2 \, dx$$

$$= \frac{1}{2A} \int_0^{\sqrt{2}} (x^4 + 2x^2 + 1) \, dx$$

$$= \left(\frac{1}{2A}\right) \left[\frac{x^5}{5} + (2)\left(\frac{x^3}{3}\right) + x\right]_0^{\sqrt{2}}$$

$$= \left(\frac{1}{2A}\right) \left[\frac{(\sqrt{2})^5}{5} + 2\frac{(\sqrt{2})^3}{3} + \sqrt{2}\right]$$

$$= \left(\frac{1}{2A}\right) \left[\left(\frac{47}{15}\right)\sqrt{2}\right]$$

$$= \left(\frac{1}{2}\right)\left(\frac{3}{5\sqrt{2}}\right)\left[\left(\frac{47}{15}\right)\sqrt{2}\right]$$

$$= \frac{47}{50}$$

To find the moments of inertia, use Eq. 3.7.

$$I_x = \frac{1}{3} \int_0^{\sqrt{2}} (x^2 + 1)^3 \, dx$$

$$= \frac{1}{3} \int_0^{\sqrt{2}} (x^6 + 3x^4 + 3x^2 + 1) \, dx$$

$$= \left(\frac{1}{3}\right) \left[\frac{x^7}{7} + \frac{3x^5}{5} + x^3 + x\right]_0^{\sqrt{2}}$$

$$= \left(\frac{1}{3}\right) \left[\frac{(\sqrt{2})^7}{7} + \frac{(3)(\sqrt{2})^5}{5} + (\sqrt{2})^3 + \sqrt{2}\right]$$

$$= \left(\frac{229}{105}\right)\sqrt{2}$$

$$I_y = \int_0^{\sqrt{2}} x^2(x^2 + 1) \, dx$$

$$= \int_0^{\sqrt{2}} (x^4 + x^2) \, dx$$

$$= \left[\frac{x^5}{5} + \frac{x^3}{3}\right]_0^{\sqrt{2}}$$

$$= \frac{(\sqrt{2})^5}{5} + \frac{(\sqrt{2})^3}{3}$$

$$= \left(\frac{22}{15}\right)\sqrt{2}$$

(b) Proceed as in Exs. 3.3 and 3.4. From Eq. 3.5 with $c = 0$, $d = 2$, and $x = g(y) = y$ (see the figure in part (b) of problem),

$$A = \int_0^2 y \, dy$$

$$= \left[\frac{y^2}{2}\right]_0^2 = 2$$

The centroid is located at (x_c, y_c).

$$x_c = \frac{1}{2A} \int_0^2 y^2 \, dy$$

$$= \left(\frac{1}{2A}\right) \left[\frac{y^3}{3}\right]_0^2$$

$$= \left[\frac{1}{(2)(2)}\right]\left(\frac{8}{3}\right)$$

$$= \frac{2}{3}$$

$$y_c = \frac{1}{A} \int_0^2 (y)(y) \, dy$$

$$= \frac{1}{A} \int_0^2 y^2 \, dy$$

$$= \left(\frac{1}{A}\right) \left[\frac{y^3}{3}\right]_0^2$$

$$= \left(\frac{1}{A}\right)\left(\frac{8}{3}\right)$$

$$= \left(\frac{1}{2}\right)\left(\frac{8}{3}\right)$$

$$= \frac{4}{3}$$

To find the moments of inertia, use Eq. 3.8.

$$I_x = \int_0^2 y^2(y) \, dy = \int_0^2 y^3 \, dy$$

$$= \left[\frac{y^4}{4}\right]_0^2 = 4$$

$$I_y = \frac{1}{3} \int_0^2 y^3 \, dy = \frac{4}{3}$$

Chapter Four

1. (a) This is a homogeneous first-order linear differential equation with constant coefficients. The characteristic equation is

$$r + 3 = 0$$

$$r = -3$$

Use Eq. 4.14 with $a = 3$. The general solution is given by

$$y(x) = Ce^{-3x} \quad [C \text{ is an arbitrary constant.}]$$

(b) This is a homogeneous first-order linear differential equation with constant coefficients. The characteristic equation is

$$r - \frac{1}{2} = 0$$

$$r = \frac{1}{2}$$

Use Eq. 4.14 with $a = -(1/2)$. The general solution is given by

$$y(x) = Ce^{\frac{x}{2}} \quad [C \text{ is an arbitrary constant.}]$$

(c) This is a homogeneous first-order linear differential equation with constant coefficients. The characteristic equation is

$$r - \frac{2}{5} = 0$$
$$r = \frac{2}{5}$$

Use Eq. 4.14 with $a = -(2/5)$. The general solution is given by

$$y(x) = Ce^{\frac{2x}{5}} \quad [C \text{ is an arbitrary constant.}]$$

2. (a) This is a homogeneous second-order linear differential equation with constant coefficients. The characteristic equation is

$$r^2 - r = 0$$
$$r(r - 1) = 0$$
$$r = 0, 1$$

Use Eq. 4.16 with $r_1 = 0$ and $r_2 = 1$. The general solution is given by

$$y(x) = c_1 e^{0x} + c_2 e^x$$
$$= c_1 + c_2 e^x \quad [c_1 \text{ and } c_2 \text{ are arbitrary constants.}]$$

(b) This is a homogeneous second-order linear differential equation with constant coefficients. The characteristic equation is

$$r^2 - 9 = 0$$
$$(r - 3)(r + 3) = 0 \quad [\text{see App. 3}]$$
$$r = \pm 3$$

Use Eq. 4.16 with $r_1 = -3$ and $r_2 = 3$. The general solution is given by

$$y(x) = c_1 e^{-3x} + c_2 e^{3x} \quad \begin{bmatrix} c_1 \text{ and } c_2 \text{ are} \\ \text{arbitrary constants.} \end{bmatrix}$$

(c) This is a homogeneous second-order linear differential equation with constant coefficients. The characteristic equation is

$$r^2 - 2r - 3 = 0$$
$$(r - 3)(r + 1) = 0$$
$$r = -1, 3$$

Use Eq. 4.16 with $r_1 = -1$ and $r_2 = 3$. The general solution is given by

$$y(x) = c_1 e^{-x} + c_2 e^{3x} \quad [c_1 \text{ and } c_2 \text{ are arbitrary constants.}]$$

3. (a) This is a homogeneous second-order linear differential equation with constant coefficients. The characteristic equation is

$$r^2 - 16r + 64 = 0$$
$$(r - 8)^2 = 0$$
$$r = 8, 8$$

Use Eq. 4.17 with $a = -8$. The general solution is given by

$$y(x) = (c_1 + c_2 x)e^{8x} \quad [c_1 \text{ and } c_2 \text{ are arbitrary constants.}]$$

(b) This is a homogeneous second-order linear differential equation with constant coefficients.

$$5y'' + 50y' + 125y = 0$$
$$y'' + 10y' + 25y = 0$$

The characteristic equation is

$$r^2 + 10r + 25 = 0$$
$$(r + 5)^2 = 0$$
$$r = -5, -5$$

Use Eq. 4.17 with $a = 5$. The general solution is given by

$$y(x) = (c_1 + c_2 x)e^{-5x} \quad \begin{bmatrix} c_1 \text{ and } c_2 \text{ are} \\ \text{arbitrary constants.} \end{bmatrix}$$

4. (a) This is a homogeneous second-order linear differential equation with constant coefficients. The characteristic equation is

$$r^2 + r + 1 = 0$$
$$r = \frac{-1 \pm \sqrt{1 - (4)(1)(1)}}{2}$$
$$= \frac{-1 \pm i\sqrt{3}}{2}$$

Use Eq. 4.18 with $\alpha = -(1/2)$ and $\beta = \sqrt{3}/2$. The general solution is given by

$$y(x) = e^{-\frac{x}{2}}\left[c_1 \cos\left(\frac{\sqrt{3}x}{2}\right) + c_2 \sin\left(\frac{\sqrt{3}x}{2}\right)\right]$$

$$[c_1 \text{ and } c_2 \text{ are arbitrary constants.}]$$

(b) This is a homogeneous second-order linear differential equation with constant coefficients. The characteristic equation is

$$r^2 - 4r + 5 = 0$$

$$r = \frac{4 \pm \sqrt{16 - (4)(1)(5)}}{2}$$

$$= \frac{4 \pm 2i}{2} = 2 \pm i$$

Use Eq. 4.18 with $\alpha = 2$ and $\beta = 1$. The general solution is given by

$$y(x) = e^{2x}(c_1 \cos x + c_2 \sin x) \qquad \left[\begin{array}{l} c_1 \text{ and } c_2 \text{ are} \\ \text{arbitrary constants.} \end{array}\right]$$

(c) This is a homogeneous second-order linear differential equation with constant coefficients. The characteristic equation is

$$r^2 + 16 = 0$$

$$r^2 = -16 = 16i^2$$

$$r = \pm 4i$$

Use Eq. 4.18 with $\alpha = 0$ and $\beta = 4$. The general solution is given by

$$y(x) = e^{0x}(c_1 \cos 4x + c_2 \sin 4x)$$

$$= c_1 \cos 4x + c_2 \sin 4x \qquad \left[\begin{array}{l} c_1 \text{ and } c_2 \text{ are} \\ \text{arbitrary constants.} \end{array}\right]$$

5. (a) First find the general solution of the differential equation. This is a homogeneous first-order linear differential equation with constant coefficients. The characteristic equation is

$$r - 5 = 0$$

$$r = 5$$

Use Eq. 4.14 with $a = -5$. The general solution is given by

$$y(t) = Ce^{5t} \quad [C \text{ is an arbitrary constant.}]$$

Use the initial condition to evaluate the constant C.

$$y(0) = 3$$

$$Ce^{(5)(0)} = 3$$

$$C(1) = 3$$

$$C = 3$$

The particular, or specific, solution is given by

$$y(t) = 3e^{5t}$$

6. First find the general solution of the differential equation. This is a homogeneous second-order linear differential equation with constant coefficients. The characteristic equation is

$$r^2 + 4 = 0$$

$$r^2 = -4 = 4i^2$$

$$r = \pm 2i$$

Use Eq. 4.18 with $\alpha = 0$ and $\beta = 2$. The general solution is given by

$$y(x) = c_1 \cos 2x + c_2 \sin 2x \qquad \left[\begin{array}{l} c_1 \text{ and } c_2 \text{ are} \\ \text{arbitrary constants.} \end{array}\right]$$

Use the boundary conditions to evaluate the constants c_1 and c_2.

$$y(0) = 0$$

$$c_1 \cos 2(0) + c_2 \sin 2(0) = 0$$

$$c_1 + c_2(0) = 0$$

$$c_1 = 0$$

$$y\left(\frac{\pi}{4}\right) = 1$$

$$c_2 \sin 2\left(\frac{\pi}{4}\right) = 1$$

$$c_2 \sin \frac{\pi}{2} = 1$$

$$c_2(1) = 1$$

$$c_2 = 1$$

The particular, or specific, solution is given by

$$y(x) = c_1 \cos 2x + c_2 \sin 2x$$

$$= 0 + 1 \sin 2x$$

$$= \sin 2x$$

7. This is a homogeneous second-order linear differential equation with constant coefficients. The characteristic equation is given by

$$r^5 + r^4 - r^3 - 3r^2 + 2 = 0$$

$$(r + 1)(r^2 + 2r + 2)(r - 1)^2 = 0$$

This has solutions

$$r = -1, -1 \pm i, 1, 1$$

As in Ex. 4.14, using Eqs. 4.8 through 4.10, the general solution is given by

$$y(x) = c_1 e^{-x} + e^{-x}(c_2 \cos x + c_3 \sin x) + (c_4 + c_5 x)e^{x}$$

$$[c_1, \ldots, c_5 \text{ are arbitrary constants.}]$$

8. The general or complete solution is the sum of the complementary and particular solutions.

(a)
$$2y'' + 2y' - 4y = 10e^{3x}$$
$$y'' + y' - 2y = 5e^{3x}$$

Complementary solution:
The homogeneous equation is

$$y'' + y' - 2y = 0$$

The characteristic equation is

$$r^2 + r - 2 = 0$$
$$(r + 2)(r - 1) = 0$$
$$r_1 = -2$$
$$r_2 = 1$$

From Eq. 4.16, the complementary solution is given by

$$y_h(x) = c_1 e^{-2x} + c_2 e^x \qquad \begin{bmatrix} c_1 \text{ and } c_2 \text{ are} \\ \text{arbitrary constants.} \end{bmatrix}$$

Particular solution:
Assume the particular solution is of the form $x^s A e^{3x}$ (where A is constant) since that is the form of the forcing function in the nonhomogeneous equation. Let $s = 0$ and $y_p(x) = A e^{3x}$. No combination of the constants c_1 and c_2 in y_h can generate this y_p. Consequently, $y_p(x) = A e^{3x}$ does not solve the homogeneous equation and is the correct suggestion for the particular solution $y_p(x)$. The first and second derivatives are

$$y_p'(x) = 3A e^{3x}$$
$$y_p''(x) = 9A e^{3x}$$
$$y_p'' + y_p' - 2y_p = 5e^{3x}$$
$$9A e^{3x} + 3A e^{3x} - 2A e^{3x} = 5e^{3x}$$
$$10A e^{3x} = 5e^{3x}$$
$$A = \frac{1}{2}$$
$$y_p(x) = \frac{1}{2} e^{3x}$$

Complete solution:

$$y(x) = y_h(x) + y_p(x)$$
$$= c_1 e^{-2x} + c_2 e^x + \frac{1}{2} e^{3x} \qquad \begin{bmatrix} c_1 \text{ and } c_2 \text{ are} \\ \text{arbitrary constants.} \end{bmatrix}$$

(b)
$$4y'' - 4y' + y = \sin x$$

Complementary solution:
The homogeneous equation is

$$4y'' - 4y' + y = 0$$

The characteristic equation is

$$4r^2 - 4r + 1 = 0$$
$$(2r - 1)^2 = 0$$
$$r = \frac{1}{2}, \frac{1}{2}$$

From Eq. 4.17, the complementary solution is given by

$$y_h(x) = (c_1 + c_2 x) e^{\frac{1}{2}x} \qquad \begin{bmatrix} c_1 \text{ and } c_2 \text{ are} \\ \text{arbitrary constants.} \end{bmatrix}$$

Particular solution:
Assume the particular solution is of the form $x^s(A \sin x + B \cos x)$ (where A and B are constant) since that is the form of the forcing function in the nonhomogeneous equation. Let $s = 0$ and $y_p(x) = A \sin x + B \cos x$. No combination of the constants c_1 and c_2 in y_h can generate any of the terms in this y_p. Consequently, no term in $y_p(x) = A \sin x + B \cos x$ solves the homogeneous equation. This y_p is therefore the correct suggestion for the particular solution. The first and second derivatives are

$$y_p'(x) = A \cos x - B \sin x$$
$$y_p''(x) = -A \sin x - B \cos x$$
$$4y_p'' - 4y_p' + y_p = \sin x$$
$$(4)(-A \sin x - B \cos x)$$
$$-(4)(A \cos x - B \sin x)$$
$$+(A \sin x + B \cos x) = \sin x$$
$$(-3A + 4B) \sin x$$
$$+(-4A - 3B) \cos x = \sin x$$

Equate the coefficients of $\sin x$ and $\cos x$.

$$-3A + 4B = 1$$
$$-4A - 3B = 0$$

Solving this system leads to $A = -(3/25)$, $B = 4/25$, and

$$y_p(x) = \left(\frac{1}{25}\right)(-3 \sin x + 4 \cos x)$$

Complete solution:

$$y(x) = y_h(x) + y_p(x)$$
$$= (c_1 + c_2 x) e^{\frac{1}{2}x} + \left(\frac{1}{25}\right)(-3 \sin x + 4 \cos x)$$

$$[c_1 \text{ and } c_2 \text{ are arbitrary constants.}]$$

(c)

$$y'' + y' + y = 2x$$

Complementary solution:
The homogeneous equation is

$$y'' + y' + y = 0$$

The characteristic equation is

$$r^2 + r + 1 = 0$$

$$r = \frac{-1 \pm \sqrt{1 - (4)(1)(1)}}{2}$$

$$= \frac{-1 \pm i\sqrt{3}}{2}$$

From Eq. 4.18 with $\alpha = -(1/2)$ and $\beta = \sqrt{3}/2$, the complementary solution is given by

$$y_h(x) = e^{-\frac{x}{2}} \left[c_1 \cos\left(\frac{\sqrt{3}x}{2}\right) + c_2 \sin\left(\frac{\sqrt{3}x}{2}\right) \right]$$

$$[c_1 \text{ and } c_2 \text{ are arbitrary constants.}]$$

Particular solution:
Assume the particular solution is of the form $x^s(Ax+B)$ (where A and B are constant) since that is the form of the forcing function in the nonhomogeneous equation. Let $s = 0$ and $y_p(x) = Ax + B$. No combination of the constants c_1 and c_2 in y_h can generate any term in this y_p. Consequently, no term in $y_p(x) = Ax + B$ solves the homogeneous equation. This y_p is therefore the correct suggestion for the particular solution. The first and second derivatives are

$$y_p'(x) = A$$
$$y_p''(x) = 0$$
$$y_p'' + y_p' + y_p = 2x$$
$$0 + A + Ax + B = 2x$$
$$(A + B) + Ax = 2x$$

Equate coefficients of terms with different powers of x.

$$A + B = 0$$
$$A = 2$$

Solving this system leads to $A = 2$, $B = -2$, and

$$y_p(x) = (2)(x - 1)$$

Complete solution:

$$y(x) = y_h(x) + y_p(x)$$
$$= e^{-\frac{x}{2}} \left[c_1 \cos\left(\frac{\sqrt{3}x}{2}\right) + c_2 \sin\left(\frac{\sqrt{3}x}{2}\right) \right]$$
$$+ (2)(x - 1) \quad \left[\begin{array}{c} c_1 \text{ and } c_2 \text{ are} \\ \text{arbitrary constants.} \end{array}\right]$$

(d)

$$y'' + y' = 4e^{-x}$$

Complementary solution:
The homogeneous equation is

$$y'' + y' = 0$$

The characteristic equation is

$$r^2 + r = 0$$
$$r(r + 1) = 0$$
$$r_1 = -1$$
$$r_2 = 0$$

From Eq. 4.16, the complementary solution is given by

$$y_h(x) = c_1 e^{-x} + c_2 e^{0x}$$
$$= c_1 e^{-x} + c_2 \quad [c_1 \text{ and } c_2 \text{ are arbitrary constants.}]$$

Particular solution:
Assume the particular solution is of the form $x^s A e^{-x}$ (where A is constant) since that is the form of the forcing function in the nonhomogeneous equation. Let $s = 0$ and $y_p(x) = Ae^{-x}$. If $c_1 = A$ and $c_2 = 0$ in y_h, $y_p(x) = Ae^{-x}$ is seen to be part of y_h and therefore solves the homogeneous equation. Try $s = 1$ so that $y_p(x) = xAe^{-x}$. No choice of the constants c_1 and c_2 in y_h will generate this y_p. Consequently, $y_p(x) = Axe^{-x}$ does not solve the homogeneous equation and is therefore the correct suggestion for the particular solution $y_p(x)$. The first and second derivatives are

$$y_p'(x) = -xAe^{-x} + Ae^{-x}$$
$$y_p''(x) = -Ae^{-x} - (Ae^{-x} - xAe^{-x})$$
$$= -2Ae^{-x} + xAe^{-x}$$
$$y_p'' + y_p' = 4e^{-x}$$
$$-2Ae^{-x} + xAe^{-x}$$
$$+(-xAe^{-x} + Ae^{-x}) = 4e^{-x}$$
$$-Ae^{-x} = 4e^{-x}$$
$$-A = 4$$
$$A = -4$$
$$y_p(x) = -4xe^{-x}$$

Complete solution:

$$y(x) = y_h(x) + y_p(x)$$
$$= c_1 e^{-x} + c_2 - 4xe^{-x} \quad \left[\begin{array}{l} c_1 \text{ and } c_2 \text{ are} \\ \text{arbitrary constants.} \end{array}\right]$$

9. (a) First find the complete solution of the differential equation.

$$2\frac{dy}{dt} - 3y = \cos t$$

Complementary solution:
The homogeneous equation is

$$2\frac{dy}{dt} - 3y = 0$$

The characteristic equation is

$$2r - 3 = 0$$
$$r = \frac{3}{2}$$

The complementary solution is

$$y_h(t) = Ce^{\frac{3}{2}t} \quad [C \text{ is an arbitrary constant.}]$$

Particular solution:
Suggest $y_p(t) = t^s(A\sin t + B\cos t)$ (where A and B are constants) since the forcing function takes this form. Let $s = 0$ so that $y_p(t) = A\sin t + B\cos t$. No value of the constant C will lead to any of the terms in the suggested y_p. It follows that neither of the terms in $y_p(t) = A\sin t + B\cos t$ solves the homogeneous equation. This y_p is therefore the correct form of the particular solution. The first derivative is

$$\frac{dy_p}{dt} = A\cos t - B\sin t$$
$$2\frac{dy_p}{dt} - 3y_p = \cos t$$
$$(2)(A\cos t - B\sin t)$$
$$-(3)(A\sin t + B\cos t) = \cos t$$
$$(-2B - 3A)\sin t + (2A - 3B)\cos t = \cos t$$

Equate the coefficients of $\sin t$ and $\cos t$.

$$-2B - 3A = 0$$
$$2A - 3B = 1$$

Solve this system to obtain $A = 2/13$ and $B = -(3/13)$.

$$y_p(t) = \left(\frac{1}{13}\right)(2\sin t - 3\cos t)$$

Complete solution:

$$y(t) = y_h(t) + y_p(t)$$
$$= Ce^{\frac{3}{2}t} + \left(\frac{1}{13}\right)(2\sin t - 3\cos t)$$

Use the initial condition to evaluate C.

$$y(0) = C - \frac{3}{13} = 0$$
$$C = \frac{3}{13}$$

The specific solution is

$$y(t) = \left(\frac{1}{13}\right)(3e^{\frac{3}{2}t} + 2\sin t - 3\cos t)$$

(b) First find the complete solution of the differential equation.

$$\frac{dy}{dt} + 8y = 3e^{-8t}$$

Complementary solution:
The homogeneous equation is

$$\frac{dy}{dt} + 8y = 0$$

The characteristic equation is

$$r + 8 = 0$$
$$r = -8$$

The complementary solution is

$$y_h(t) = Ce^{-8t} \quad [C \text{ is an arbitrary constant.}]$$

Particular solution:
Suggest $y_p(t) = t^s Ae^{-8t}$ (where A is a constant) since the forcing function takes this form. Let $s = 0$ so that $y_p(t) = Ae^{-8t}$. Choosing $C = A$ in y_h will generate this y_p. Consequently, this y_p solves the homogeneous equation. Try $s = 1$ so that $y_p(t) = Ate^{-8t}$. No value of the constant C in y_h will generate this y_p. It follows that $y_p(t) = Ate^{-8t}$ does not solve the homogeneous equation. This y_p is therefore the correct form of the particular solution. The first derivative is

$$\frac{dy_p}{dt} = Ae^{-8t} - 8Ate^{-8t}$$
$$\frac{dy_p}{dt} + 8y_p = 3e^{-8t}$$
$$Ae^{-8t} - 8Ate^{-8t} + 8Ate^{-8t} = 3e^{-8t}$$
$$Ae^{-8t} = 3e^{-8t}$$
$$A = 3$$
$$y_p(t) = 3te^{-8t}$$

Complete solution:

$$y(t) = y_h(t) + y_p(t)$$
$$= Ce^{-8t} + 3te^{-8t}$$

Use the initial condition to evaluate C.

$$y(0) = C + 0 = 1$$
$$C = 1$$

The specific solution is

$$y(t) = e^{-8t} + 3te^{-8t}$$

(c) First find the complete solution of the differential equation.

$$\frac{d^2y}{dt^2} + 2\frac{dy}{dt} + y = e^{-t}$$

Complementary solution:
The homogeneous equation is

$$\frac{d^2y}{dt^2} + 2\frac{dy}{dt} + y = 0$$

The characteristic equation is

$$r^2 + 2r + 1 = 0$$
$$(r + 1)^2 = 0$$
$$r = -1, -1$$

From Eq. 4.17, the complementary solution is

$$y_h(t) = (c_1 + c_2t)e^{-t} \quad [c_1 \text{ and } c_2 \text{ are arbitrary constants.}]$$

Particular solution:
Suggest $y_p(t) = t^s A e^{-t}$ (where A is a constant) since the forcing function takes this form. Let $s = 0$ so that $y_p(t) = Ae^{-t}$. Choosing $c_1 = A$ and $c_2 = 0$ in y_h generates this y_p. Consequently, this y_p solves the homogeneous equation. Try $s = 1$ so that $y_p(t) = Ate^{-t}$. Choosing $c_1 = 0$ and $c_2 = A$ in y_h generates this y_p. Consequently, this y_p also solves the homogeneous equation. Try $s = 2$ so that $y_p(t) = At^2e^{-t}$. In this case, no choice of constants c_1 and c_2 in y_h will generate this y_p. It follows that $y_p(t) = At^2e^{-t}$ does not solve the homogeneous equation. This y_p is therefore the correct form of the particular solution. The first and second derivatives are

$$\frac{dy_p}{dt} = -At^2e^{-t} + 2Ate^{-t}$$

$$\frac{d^2y_p}{dt^2} = At^2e^{-t} - 2Ate^{-t} + 2Ae^{-t} - 2Ate^{-t}$$
$$= At^2e^{-t} - 4Ate^{-t} + 2Ae^{-t}$$

Substitute into the differential equation.

$$\frac{d^2y_p}{dt^2} + 2\frac{dy_p}{dt} + y_p = e^{-t}$$
$$At^2e^{-t} - 4Ate^{-t} + 2Ae^{-t}$$
$$+2(-At^2e^{-t} + 2Ate^{-t}) + At^2e^{-t} = e^{-t}$$
$$2Ae^{-t} = e^{-t}$$
$$A = \frac{1}{2}$$
$$y_p(t) = \frac{1}{2}t^2e^{-t}$$

Complete solution:

$$y(t) = y_h(t) + y_p(t)$$
$$= (c_1 + c_2t)e^{-t} + \frac{1}{2}t^2e^{-t} \quad \begin{bmatrix} c_1 \text{ and } c_2 \text{ are} \\ \text{arbitrary constants.} \end{bmatrix}$$

Use the initial conditions to evaluate c_1 and c_2.

$$y(0) = c_1 = 0$$
$$y(t) = c_2te^{-t} + \frac{1}{2}t^2e^{-t}$$
$$\frac{dy}{dt} = -c_2te^{-t} + c_2e^{-t} - \frac{1}{2}t^2e^{-t} + te^{-t}$$
$$\left(\frac{dy}{dt}\right)(0) = c_2 = 1$$

The specific solution is

$$y(t) = te^{-t} + \frac{1}{2}t^2e^{-t}$$

10. First find the complete solution of the differential equation.

$$y' + 2y = \cos x + \sin x$$

Complementary solution:
The homogeneous equation is

$$y' + 2y = 0$$

The characteristic equation is

$$r + 2 = 0$$
$$r = -2$$

The complementary solution is

$$y_h(x) = Ce^{-2x} \quad [C \text{ is an arbitrary constant.}]$$

Particular solution:
Suggest $y_p(x) = x^s(A\sin x + B\cos x)$ (where A and B are constants) since the forcing function takes this form. Let $s = 0$ so that $y_p(x) = A\sin x + B\cos x$. No value of the constant C will lead to any of the terms in the suggested y_p. It follows that neither of the terms in $y_p(x) = A\sin x + B\cos x$ solves the homogeneous equation. This y_p is therefore the correct form of the particular solution. The first derivative is

$$y' = A\cos x - B\sin x$$
$$y_p' + 2y_p = \cos x + \sin x$$
$$(A\cos x - B\sin x)$$
$$+(2)(A\sin x + B\cos x) = \cos x + \sin x$$
$$(A + 2B)(\cos x) + (2A - B)(\sin x) = \cos x + \sin x$$

Equate coefficients of $\sin x$ and $\cos x$.

$$A + 2B = 1$$
$$2A - B = 1$$

Solve this system to obtain $A = 3/5$ and $B = 1/5$.

$$y_p(x) = \left(\frac{1}{5}\right)(3\sin x + \cos x)$$

Complete solution:

$$y(x) = y_h(x) + y_p(x)$$
$$= Ce^{-2x} + \left(\frac{1}{5}\right)(3\sin x + \cos x)$$

Use the boundary condition to evaluate C.

$$y(0) = C + \frac{1}{5} = 0$$
$$C = -\frac{1}{5}$$

The specific solution is

$$y(x) = \left(\frac{1}{5}\right)[-e^{-2x} + (3\sin x + \cos x)]$$

11. First find the complete solution of the differential equation.

$$6y'' = 7e^{-2x} + y - y'$$
$$6y'' + y' - y = 7e^{-2x}$$

Complementary solution:
The homogeneous equation is

$$6y'' + y' - y = 0$$

The characteristic equation is

$$6r^2 + r - 1 = 0$$
$$(3r - 1)(2r + 1) = 0$$
$$r = -\frac{1}{2}, \frac{1}{3}$$

From Eq. 4.16, the complementary solution is

$$y_h(x) = c_1 e^{-\frac{x}{2}} + c_2 e^{\frac{x}{3}} \quad \left[\begin{array}{l} c_1 \text{ and } c_2 \text{ are} \\ \text{arbitrary constants.} \end{array}\right]$$

Particular solution:
Suggest $y_p(x) = x^s Ae^{-2x}$ (where A is a constant) since the forcing function takes this form. Let $s = 0$ so that $y_p(x) = Ae^{-2x}$. No choice of the constants c_1 and c_2 in y_h will generate this y_p. It follows that $y_p(x) = Ae^{-2x}$ does not solve the homogeneous equation. This y_p is therefore the correct form of the particular solution. The first and second derivatives are

$$y_p' = -2Ae^{-2x}$$
$$y_p'' = 4Ae^{-2x}$$

Substitute into the differential equation.

$$6y_p'' + y_p' - y_p = 7e^{-2x}$$
$$24Ae^{-2x} - 2Ae^{-2x} - Ae^{-2x} = 7e^{-2x}$$
$$21Ae^{-2x} = 7e^{-2x}$$
$$A = \frac{1}{3}$$
$$y_p(x) = \frac{1}{3}e^{-2x}$$

Complete solution:

$$y(x) = y_h(x) + y_p(x)$$
$$= c_1 e^{-\frac{x}{2}} + c_2 e^{\frac{x}{3}} + \frac{1}{3}e^{-2x} \quad \left[\begin{array}{l} c_1 \text{ and } c_2 \text{ are} \\ \text{arbitrary constants.} \end{array}\right]$$

Use the boundary conditions to evaluate c_1 and c_2.

$$y(0) = c_1 + c_2 + \frac{1}{3} = 0$$
$$y' = -\left(\frac{c_1}{2}\right)e^{-\frac{x}{2}} + \left(\frac{c_2}{3}\right)e^{\frac{x}{3}} - \left(\frac{2}{3}\right)e^{-2x}$$
$$y'(0) = -\left(\frac{c_1}{2}\right) + \frac{c_2}{3} - \frac{2}{3} = 1$$
$$c_1 + c_2 = -\frac{1}{3}$$
$$-\frac{c_1}{2} + \frac{c_2}{3} = \frac{5}{3}$$

Solving this system for c_1 and c_2 leads to $c_1 = -(32/15)$ and $c_2 = 9/5$.

The specific solution is

$$y(x) = -\left(\frac{32}{15}\right)e^{-\frac{x}{2}} + \left(\frac{9}{5}\right)e^{\frac{x}{3}} + \left(\frac{1}{3}\right)e^{-2x}$$

Appendix

2

Solutions to FE-Style Exam Problems

Chapter One

1.

$$y = e^{-x} \ln 2x$$

$$\frac{dy}{dx} = e^{-x}\left(\frac{2}{2x}\right) + \ln 2x\left(-e^{-x}\right)$$

$$= e^{-x}\left(\frac{1}{x} - \ln 2x\right)$$

Answer is (B).

2. Use l'Hôpital's rule. Let $x = 1$ in $(3x^2 + 2x - 5)/(x^4 + 3x^2 - 4)$. The limit is indeterminate of type $\frac{0}{0}$.

$$\lim_{x \to 1} \frac{3x^2 + 2x - 5}{x^4 + 3x^2 - 4} = \lim_{x \to 1} \frac{6x + 2}{4x^3 + 6x}$$

$$= \frac{8}{10}$$

$$= \frac{4}{5}$$

Answer is (C).

3.

$$y = \frac{1}{x}$$

$$y' = -\frac{1}{x^2}$$

$$y'' = \frac{2}{x^3}$$

$$K = \left| \frac{y''}{\left(1 + (y')^2\right)^{\frac{3}{2}}} \right| = \left| \frac{\frac{2}{x^3}}{\left(1 + \frac{1}{x^4}\right)^{\frac{3}{2}}} \right|$$

Let $x = 1$.

$$K = \left| \frac{2}{2^{\frac{3}{2}}} \right| = \frac{1}{\sqrt{2}}$$

Answer is (D).

4.

$$f(t) = t^3 + 3t^2 - 9t + 5$$

$$f'(t) = 3t^2 + 6t - 9$$

$$= (3)(t^2 + 2t - 3)$$

$$= (3)(t + 3)(t - 1)$$

It follows that $f'(t) = 0$ when $t = -3, 1$ and that the critical points are located at $t = 3$ and $t = 1$.

$$f''(t) = 6t + 6$$

$$f''(-3) = -12 < 0$$

$$f''(1) = 12 > 0$$

There is a local maximum at $t = -3$, and a local minimum at $t = 1$. Compare the values of $f(t)$ at the endpoints with those at the local extrema.

$$f(-3) = (-3)^3 + (3)(-3)^2 - (9)(-3) + 5 = 32$$

$$f(1) = (1)^3 + (3)(1)^2 - (9)(1) + 5 = 0$$

$$f(-6) = (-6)^3 + (3)(-6)^2 - (9)(-6) + 5 = -49$$

$$f(2.5) = (2.5)^3 + (3)(2.5)^2 - (9)(2.5) + 5 = 16.9$$

The maximum value is 32, occurring at the critical point $t = -3$. The minimum value is -49, occurring at the endpoint $t = -6$.

Answer is (A).

5.

$$\varphi(x, y, z) = \frac{-1}{\sqrt{x^2 + y^2 + z^2}} = -\left(x^2 + y^2 + z^2\right)^{-\frac{1}{2}}$$

$$\frac{\partial \varphi}{\partial x} = \left(\frac{1}{2}\right)\left(x^2 + y^2 + z^2\right)^{-\frac{3}{2}}(2x)$$

$$= \frac{x}{\left(x^2 + y^2 + z^2\right)^{\frac{3}{2}}}$$

$$\frac{\partial^2 \varphi}{\partial x^2} = \frac{\left(x^2 + y^2 + z^2\right)^{\frac{3}{2}}(1)}{-(x)\left(\frac{3}{2}\right)\left(x^2 + y^2 + z^2\right)^{\frac{1}{2}}(2x)}{\left(x^2 + y^2 + z^2\right)^3}$$

$$= \frac{\left(x^2 + y^2 + z^2\right)^{\frac{1}{2}}\left[\left(x^2 + y^2 + z^2\right) - 3x^2\right]}{\left(x^2 + y^2 + z^2\right)^3}$$

$$= \frac{-2x^2 + y^2 + z^2}{\left(x^2 + y^2 + z^2\right)^{\frac{5}{2}}}$$

By the symmetry of φ (or by direct calculation),

$$\frac{\partial^2 \varphi}{\partial y^2} = \frac{-2y^2 + x^2 + z^2}{\left(x^2 + y^2 + z^2\right)^{\frac{5}{2}}}$$

$$\frac{\partial^2 \varphi}{\partial z^2} = \frac{-2z^2 + y^2 + x^2}{\left(x^2 + y^2 + z^2\right)^{\frac{5}{2}}}$$

It follows that

$$\frac{\partial^2 \varphi}{\partial x^2} + \frac{\partial^2 \varphi}{\partial y^2} + \frac{\partial^2 \varphi}{\partial z^2} = \frac{\begin{array}{c}(-2)(x^2 + y^2 + z^2) \\ + (2)(x^2 + y^2 + z^2)\end{array}}{\left(x^2 + y^2 + z^2\right)^{\frac{5}{2}}}$$

$$= 0$$

Answer is (A).

6. Use implicit differentiation.

$$xe^{yx} = 3$$

$$x[(y + xy')e^{yx}] + e^{yx}(1) = 0$$

$$x(y + xy') = -1$$

$$y' = \frac{-1 - xy}{x^2}$$

$$= \frac{-(xy + 1)}{x^2}$$

Answer is (C).

7.

$$f(x, y) = x^3 y + \sqrt{y}x^4 + \cos x + \tan y + \sin^6 y$$

$$\frac{\partial f}{\partial x} = 3x^2 y + 4\sqrt{y}x^3 - \sin x + 0 + 0$$

$$= 3x^2 y + 4x^3 \sqrt{y} - \sin x$$

Answer is (B).

8.

$$f(x) = ax^2 + 4x + 13$$

$$f'(x) = 2ax + 4$$

At a critical point, the first derivative, $f'(x)$, equals zero.

$$2ax + 4 = 0$$

When $x = 1$, $f'(1) = 2a + 4 = 0$ when $a = -2$. When $a = -2$, it follows that $x = 1$ is a critical point. Test to see if $a = -2$ corresponds to a local maximum at $x = 1$.

$$f''(x) = 2a$$

$$= -4 < 0$$

Consequently, $a = -2$ corresponds to a local maximum at $x = 1$.

Answer is (D).

9.

$$\lim_{x \to \infty} \frac{\ln x^{99}}{x^2} = \lim_{x \to \infty} \frac{99 \ln x}{x^2}$$

This equation is an indeterminate form of type $\frac{\infty}{\infty}$. Use l'Hôpital's rule.

$$\lim_{x \to \infty} \frac{\ln x^{99}}{x^2} = \lim_{x \to \infty} \frac{99 \ln x}{x^2}$$

$$= \lim_{x \to \infty} \frac{(99)\left(\frac{1}{x}\right)}{2x}$$

$$= \lim_{x \to \infty} \frac{99}{2x^2}$$

$$= 0$$

Answer is (C).

10.

$$x^2 + y^2 = 9$$

Use implicit differentiation to find dy/dx.

$$2x + 2yy' = 0$$

$$y' = \frac{dy}{dx} = -\frac{x}{y}$$

Answer is (A).

Chapter Two

1.

$$\int \left(\frac{x^2+x+1}{\sqrt{x}}\right) dx = \int \left(\frac{x^2+x+1}{x^{\frac{1}{2}}}\right) dx$$
$$= \int (x^{\frac{3}{2}} + x^{\frac{1}{2}} + x^{-\frac{1}{2}}) dx$$
$$= \frac{2}{5}x^{\frac{5}{2}} + \frac{2}{3}x^{\frac{3}{2}} + 2x^{\frac{1}{2}} + C$$

Answer is (B).

2.

$$A = \int_0^2 (4x^2 - 4x + 3) dx$$
$$= \left[\frac{4}{3}x^3 - 2x^2 + 3x\right]_0^2$$
$$= \frac{32}{3} - 8 + 6 - 0 = \frac{26}{3}$$

Answer is (C).

3. Let $u = \sin x$ so that $du = \cos x\, dx$. The integral becomes

$$\int \frac{du}{1+u^2} = \arctan u + C$$
$$= \arctan(\sin x) + C$$

Answer is (A).

4. Use integration by parts.

$$\int_0^{\frac{\pi}{2}} x\sin x\, dx = \left[x(-\cos x)\right]_0^{\frac{\pi}{2}} - \int_0^{\frac{\pi}{2}} (1)(-\cos x)\, dx$$
$$= 0 + \int_0^{\frac{\pi}{2}} \cos x\, dx$$
$$= \left[\sin x\right]_0^{\frac{\pi}{2}} = 1$$

Answer is (D).

5.

$$A = \int_0^{\pi} \sin x\, dx$$
$$= -\left[\cos x\right]_0^{\pi}$$
$$= -(-1-1) = 2$$

Answer is (B).

6. Use integration by parts.

$$\int (3x-1)e^x\, dx = (3x-1)e^x - \int e^x(3)\, dx$$
$$= (3x-1)e^x - 3e^x + C$$
$$= e^x(3x-4) + C$$

Answer is (A).

7. Let $u = \cos x$ so that $du = -\sin x\, dx$. The integral becomes

$$-\int \ln u\, du = -(u\ln u - u) + C$$
$$= -(\cos x\ln(\cos x) - \cos x) + C$$
$$= -\cos x[\ln(\cos x) - 1] + C$$

Answer is (C).

8.

$$\frac{4x}{(x^2-2x+1)(x+1)} = \frac{4x}{(x-1)^2(x+1)}$$
$$= \frac{A}{x-1} + \frac{B}{(x-1)^2} + \frac{C}{x+1}$$

Here, A, B, and C are constants. Clear the fractions.

$$4x = A(x-1)(x+1) + B(x+1) + C(x-1)^2$$
$$= Ax^2 - A + Bx + B + Cx^2 - 2Cx + C$$
$$= (A+C)x^2 + (B-2C)x + (B-A+C)$$

Equate the coefficients of terms with different powers of x.
$$A + C = 0$$
$$B - 2C = 4$$
$$B - A + C = 0$$

Solve using Cramer's rule or elimination to obtain $A = 1$, $B = 2$, and $C = -1$.

Alternatively, substitute $x = -1$ and 1 into the equation.

$$4x = A(x-1)(x+1) + B(x+1) + C(x-1)^2$$

This leads directly to $C = -1$ and $B = 2$. The constant A is then determined from the equation $A + C = 0$ obtained previously. Finally, $A = 1$, $B = 2$, and $C = -1$.

The integral can now be rewritten in the following simplified form.

$$\int \left(\frac{4x}{(x^2 - 2x + 1)(x + 1)} \right) dx$$

$$= \int \left(\frac{1}{x-1} + \frac{2}{(x-1)^2} - \frac{1}{x+1} \right) dx$$

$$= \ln|x-1| - \frac{2}{x-1} - \ln|x+1| + C$$

$$= \ln \left| \frac{x-1}{x+1} \right| - \frac{2}{x-1} + C$$

Answer is (D).

9.

$$\int \left(\frac{\sqrt{2}}{9 + 3x^2} \right) dx = \frac{\sqrt{2}}{3} \int \frac{dx}{3 + x^2}$$

$$= \left(\frac{\sqrt{2}}{3} \right) \left[\frac{1}{\sqrt{3}} \arctan \left(\frac{x}{\sqrt{3}} \right) \right] + C$$

$$= \frac{\sqrt{2}}{3\sqrt{3}} \arctan \left(\frac{x}{\sqrt{3}} \right) + C$$

Answer is (A).

10. Let $u = x^2 + 2x$ so that $du = (2x+2)\,dx = (2)(x+1)\,dx$. Also, when $x = -1$, $u = (-1)^2 + (2)(-1) = -1$; when $x = -1/2$, $u = (-1/2)^2 + (2)(-1/2) = -3/4$.

The integral becomes

$$\frac{1}{2} \int_{-1}^{-\frac{3}{4}} \frac{du}{u} = \left(\frac{1}{2} \right) \left[\ln|u| \right]_{-1}^{-\frac{3}{4}}$$

$$= \left(\frac{1}{2} \right) \left(\ln\left| -\frac{3}{4} \right| - \ln|-1| \right)$$

$$= \left(\frac{1}{2} \right) \left(\ln\left(\frac{3}{4} \right) - \ln(1) \right)$$

$$= \left(\frac{1}{2} \right) \left(\ln\left(\frac{3}{4} \right) - 0 \right)$$

$$= \frac{1}{2} \ln \frac{3}{4}$$

Answer is (A).

Chapter Four

1. The characteristic equation is

$$2r^2 + 32 = 0$$
$$r^2 + 16 = 0$$
$$r = \pm 4i$$

The general solution is

$$y(x) = c_1 \cos 4x + c_2 \sin 4x$$

Answer is (C).

2. First find the general solution of the differential equation. The characteristic equation is

$$2r - 6 = 0$$
$$r = 3$$

The general solution is

$$y(t) = Ce^{3t}$$

Use the initial value to find C.

$$y(0) = C = 1$$

The solution of the initial value problem is

$$y(t) = e^{3t}$$

Answer is (A).

3. First find the general solution of the differential equation. The characteristic equation is

$$r^2 + 6r + 9 = 0$$
$$(r + 3)^2 = 0$$
$$r = -3, -3$$

The general solution is

$$y(x) = (c_1 + c_2 x)e^{-3x}$$

Use the boundary conditions to find the constants c_1 and c_2.

$$y(0) = c_1 = 0$$
$$y(x) = c_2 x e^{-3x}$$
$$y'(x) = c_2(e^{-3x} - 3xe^{-3x})$$
$$y'(0) = c_2 = 1$$

The solution of the boundary value problem is

$$y(x) = xe^{-3x}$$

Answer is (D).

4. The characteristic equation is

$$r^2 - 2r + 5 = 0$$
$$r = \frac{-(-2) \pm \sqrt{(-2)^2 - (4)(1)(5)}}{2}$$
$$= 1 \pm 2i$$

The general solution is

$$y(x) = e^x(c_1 \cos 2x + c_2 \sin 2x)$$

Answer is (C).

5. First find the complementary solution, $y_h(t)$. The homogeneous equation is

$$\frac{d^2y}{dt^2} + y = 0$$

The characteristic equation is given by

$$r^2 + 1 = 0$$
$$r = \pm i$$
$$y_h(t) = c_1 \cos t + c_2 \sin t$$

For the particular solution $y_p(t)$, suggest $y_p(t) = t^s(A \cos t + B \sin t)$ since this is the form of the forcing function. Let $s = 0$ so that $y_p(t) = A \cos t + B \sin t$. Choosing $c_1 = A$ and $c_2 = B$ in y_h will generate both terms in this y_p. It follows that both terms in this y_p solve the homogeneous equation. Let $s = 1$ so that $y_p(t) = t(A \cos t + B \sin t)$. No combination of the constants c_1 and c_2 in y_h will generate this y_p. It follows that no term in this y_p solves the homogeneous equation. The correct y_p is, therefore,

$$y_p(t) = t(A \cos t + B \sin t)$$

Answer is (B).

6.
$$160y = 36y' - 2y'' + 4e^{3x}$$
$$2y'' - 36y' + 160y = 4e^{3x}$$
$$y'' - 18y' + 80y = 2e^{3x}$$

The equation is linear, second-order, and nonhomogeneous.

Answer is (A).

7. The equation is linear so that the complete or general solution is given by the sum of complementary and particular solutions.

Answer is (B).

8. First find the general solution of the differential equation. The general solution is the sum of the complementary and particular solutions.

Complementary solution:
The homogeneous equation is

$$y'' - 18y' + 80y = 0$$

The characteristic equation is

$$r^2 - 18r + 80 = 0$$
$$(r - 10)(r - 8) = 0$$
$$r = 8, 10$$

The complementary solution is

$$y_h(x) = c_1 e^{8x} + c_2 e^{10x} \qquad \begin{bmatrix} c_1 \text{ and } c_2 \text{ are} \\ \text{arbitrary constants.} \end{bmatrix}$$

Particular solution:
The forcing function, $2e^{3x}$, suggests a particular solution of the form $y_p(x) = x^s A e^{3x}$. Let $s = 0$ so that $y_p(x) = A e^{3x}$. No choice of the constants c_1 and c_2 in y_h will generate this y_p. Consequently, this y_p does not solve the homogeneous equation and is therefore the correct form of the particular solution. The first and second derivatives are

$$y_p' = 3A e^{3x}$$
$$y_p'' = 9A e^{3x}$$
$$y_p'' - 18y_p' + 80y_p = 2e^{3x}$$
$$9A e^{3x} - (18)(3A e^{3x}) + 80A e^{3x} = 2e^{3x}$$
$$35A e^{3x} = 2e^{3x}$$
$$A = \frac{2}{35}$$
$$y_p(x) = \frac{2}{35} e^{3x}$$

Complete solution:

$$y(x) = y_h(x) + y_p(x)$$
$$= c_1 e^{8x} + c_2 e^{10x} + \frac{2}{35} e^{3x}$$

Use the boundary conditions to evaluate c_1 and c_2.

$$y(0) = c_1 + c_2 + \frac{2}{35} = 0$$
$$c_1 + c_2 = -\frac{2}{35}$$
$$y'(x) = 8c_1 e^{8x} + 10c_2 e^{10x} + \frac{6}{35} e^{3x}$$
$$y'(0) = 8c_1 + 10c_2 + \frac{6}{35} = 1$$
$$8c_1 + 10c_2 = \frac{29}{35}$$

Solve the system using Cramer's rule.

$$c_1 + c_2 = -\frac{2}{35}$$
$$8c_1 + 10c_2 = \frac{29}{35}$$

The solution is $c_1 = -(7/10)$ and $c_2 = 9/14$. Finally, the complete solution is

$$y(x) = -\frac{7}{10}e^{8x} + \frac{9}{14}e^{10x} + \frac{2}{35}e^{3x}$$

Answer is (D).

9. The characteristic equation is

$$r^2 - 3r + 2 = 0$$
$$(r-2)(r-1) = 0$$
$$r = 1, 2$$

The general solution is

$$y(x) = c_1 e^x + c_2 e^{2x} \quad [c_1 \text{ and } c_2 \text{ are arbitrary constants.}]$$

Answer is (D).

10. The general solution is the sum of the complementary and particular solutions.

Complementary solution:
The homogeneous equation is

$$y'' + 4y' = 0$$

The characteristic equation is

$$r^2 + 4r = 0$$
$$r(r+4) = 0$$
$$r = -4, 0$$

The complementary solution is

$$y_h(x) = c_1 e^{-4x} + c_2 \quad [c_1 \text{ and } c_2 \text{ are arbitrary constants.}]$$

Particular solution:
The forcing function, $2e^{-4x}$, suggests a particular solution of the form $y_p(x) = x^s A e^{-4x}$. Let $s = 0$ so that $y_p(x) = A e^{-4x}$. Choosing $c_1 = A$ and $c_2 = 0$ in y_h will generate this y_p. It follows that this y_p solves the homogeneous equation. Try $s = 1$ so that $y_p(x) = x A e^{-4x}$. No choice of the constants c_1 and c_2 in y_h will generate this y_p. Consequently, this y_p does not solve the homogeneous equation and is therefore the correct form of the particular solution. The first and second derivatives are

$$y_p' = -4x A e^{-4x} + A e^{-4x}$$
$$y_p'' = -4 A e^{-4x} + 16x A e^{-4x}$$
$$\qquad - 4 A e^{-4x}$$
$$\qquad = -8 A e^{-4x} + 16x A e^{-4x}$$
$$y_p'' + 4y_p' = e^{-4x}$$
$$-8 A e^{-4x} + 16x A e^{-4x}$$
$$+ (4)(-4x A e^{-4x} + A e^{-4x}) = e^{-4x}$$
$$-4 A e^{-4x} = e^{-4x}$$
$$A = -\frac{1}{4}$$
$$y_p(x) = -\frac{x}{4} e^{-4x}$$

Complete solution:

$$y(x) = y_h(x) + y_p(x)$$
$$= c_1 e^{-4x} + c_2 - \left(\frac{x}{4}\right) e^{-4x}$$

Answer is (A).

Appendix

3

Selected Precalculus Topics

Solving Quadratic Equations

Consider the following quadratic equation.

$$ax^2 + bx + c = 0 \quad [a \neq 0] \qquad \text{(A3.1)}$$

Here, a, b, and c are real numbers. Equation A3.1 has the following solutions.

$$x = \frac{-b \pm \sqrt{b^2 - 4ac}}{2a} \qquad \text{(A3.2)}$$

There are three possible cases arising from Eq. A3.2.

Case 1:

$$b^2 - 4ac > 0$$

In this case, Eq. A3.2 gives two real and distinct solutions of Eq. A3.1.

Case 2:

$$b^2 - 4ac = 0$$

In this case, Eq. A3.2 gives two real, coincident solutions of Eq. A3.1. These solutions are given by $x = -(b/2a), -(b/2a)$.

Case 3:

$$b^2 - 4ac < 0$$

In this case, Eq. A3.2 gives two complex (nonreal) solutions of Eq. A3.1. These solutions take the following form.

$$x = \frac{-b \pm \sqrt{b^2 - 4ac}}{2a} = \frac{-b \pm i\sqrt{4ac - b^2}}{2a} \quad [i^2 = -1]$$

Factoring Formulas

To factor a polynomial is to write it as the product of two or more polynomials of lesser degree. Table A3.1 lists some of the more useful factoring formulas.

Table A3.1 Factoring Formulas

expression	factored form
$ax + ay + az$	$a(x + y + z)$
$x^2 + (a + b)x + ab$	$(x + a)(x + b)$
$x^2 + 2xy + y^2$	$(x + y)^2$
$x^2 - 2xy + y^2$	$(x - y)^2$
$x^2 - y^2$	$(x - y)(x + y)$
$x^3 + 3x^2y + 3xy^2 + y^3$	$(x + y)^3$
$x^3 - 3x^2y + 3xy^2 - y^3$	$(x - y)^3$
$x^3 + y^3$	$(x + y)(x^2 - xy + y^2)$
$x^3 - y^3$	$(x - y)(x^2 + xy + y^2)$

Here, a and b are constants [which may be complex (nonreal)].

Trigonometry

Throughout this section, the (real) argument x is measured in radians.

Trigonometric Functions

Table A3.2 Common Values of $\cos x$, $\sin x$, and $\tan x$

x	0	$\dfrac{\pi}{6}$	$\dfrac{\pi}{4}$	$\dfrac{\pi}{3}$	$\dfrac{\pi}{2}$	π	$\dfrac{3\pi}{2}$	2π
$\cos x$	1	$\dfrac{\sqrt{3}}{2}$	$\dfrac{1}{\sqrt{2}}$	$\dfrac{1}{2}$	0	-1	0	1
$\sin x$	0	$\dfrac{1}{2}$	$\dfrac{1}{\sqrt{2}}$	$\dfrac{\sqrt{3}}{2}$	1	0	-1	0
$\tan x$	0	$\dfrac{1}{\sqrt{3}}$	1	$\sqrt{3}$	∞	0	∞	0

Range

$$-1 \leq \cos x \leq 1$$
$$-1 \leq \sin x \leq 1$$
$$-\infty < \tan x < \infty$$

Periodicity

Both $\sin x$ and $\cos x$ repeat every 2π radians. That is, they are periodic with period 2π.

$$\cos (x + 2\pi) = \cos x$$
$$\sin (x + 2\pi) = \sin x$$

The function $f(x) = \tan x$ is periodic with period π.

$$\tan (x + \pi) = \tan x$$

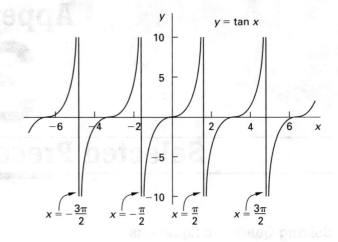

Odd/Even

$$\cos (-x) = \cos x \quad [\cos x \text{ is an even function.}]$$
$$\sin (-x) = -\sin x \quad [\sin x \text{ is an odd function.}]$$
$$\tan (-x) = -\tan x \quad [\tan x \text{ is an odd function.}]$$

In the third figure of the previous group, the lines $x = \ldots -(3\pi/2), -(\pi/2), \pi/2, 3\pi/2, \ldots$ are called vertical asymptotes. The curve tends toward (but never touches) a vertical asymptote. In this case, the vertical asymptotes occur when $\tan x = \sin x/\cos x$ becomes undefined, that is, when $\cos x = 0$.

The tangent, cotangent, secant, and cosecant of x are defined, respectively, as follows.

$$\tan x = \frac{\sin x}{\cos x}$$
$$\cot x = \frac{1}{\tan x} = \frac{\cos x}{\sin x}$$
$$\sec x = \frac{1}{\cos x}$$
$$\csc x = \frac{1}{\sin x}$$

Curves

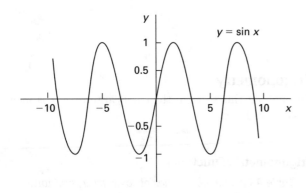

Trigonometric Identities

Let x and y be real variables, measured in radians.

$$\sin^2 x + \cos^2 x = 1$$
$$1 + \tan^2 x = \sec^2 x$$
$$1 + \cot^2 x = \csc^2 x$$
$$\cos (x \pm y) = \cos x \cos y \mp \sin x \sin y$$
$$\sin (x \pm y) = \sin x \cos y \pm \cos x \sin y$$
$$\tan (x \pm y) = \frac{\tan x \pm \tan y}{1 \mp \tan x \tan y}$$
$$\cos 2x = \cos^2 x - \sin^2 x$$
$$= 2 \cos^2 x - 1$$
$$= 1 - 2 \sin^2 x$$
$$\sin 2x = 2 \sin x \cos x$$

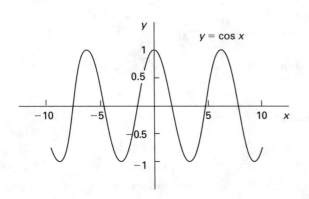

$$\sin x \sin y = \left(\frac{1}{2}\right)[\cos(x-y) - \cos(x+y)]$$

$$\sin x \cos y = \left(\frac{1}{2}\right)[\sin(x+y) + \sin(x-y)]$$

$$\cos x \cos y = \left(\frac{1}{2}\right)[\cos(x+y) + \cos(x-y)]$$

$$\sin x + \sin y = 2\sin\left(\frac{x+y}{2}\right)\cos\left(\frac{x-y}{2}\right)$$

$$\sin x - \sin y = 2\cos\left(\frac{x+y}{2}\right)\sin\left(\frac{x-y}{2}\right)$$

$$\cos x + \cos y = 2\cos\left(\frac{x+y}{2}\right)\cos\left(\frac{x-y}{2}\right)$$

$$\cos x - \cos y = 2\sin\left(\frac{x+y}{2}\right)\sin\left(\frac{y-x}{2}\right)$$

Trigonometric Limits

$$\lim_{x \to 0} \sin x = 0$$

$$\lim_{x \to 0} \cos x = 1$$

$$\lim_{x \to 0} \tan x = 0$$

$$\lim_{x \to 0} \frac{\sin x}{x} = 1$$

Exponential and Logarithmic Functions

Natural Exponential Function

The (irrational) number e (after Euler) is defined as

$$e = \lim_{m \to \infty}\left(1 + \frac{1}{m}\right)^m = 2.7182818285\ldots$$

The *natural exponential function* (so called because of its natural tendency to arise in applications) is defined by

$$y = f(x) = e^x$$

Here, x is any real number.

Properties of $f(x) = e^x$

Let x and y be real numbers.

$$y = f(x) = e^x > 0$$

$$e^0 = 1$$

$$e^{-x} = \frac{1}{e^x}$$

$$e^{x+y} = (e^x)(e^y)$$

$$e^{x-y} = (e^x)(e^{-y}) = \frac{e^x}{e^y}$$

Limiting Properties of $f(x) = e^x$

$$\lim_{x \to \infty} e^x = \infty$$

$$\lim_{x \to -\infty} e^x = 0$$

$$\lim_{x \to \infty} e^{-x} = 0$$

$$\lim_{x \to -\infty} e^{-x} = \infty$$

Curves

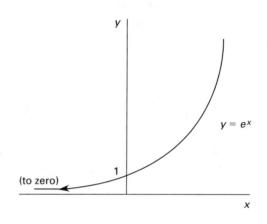

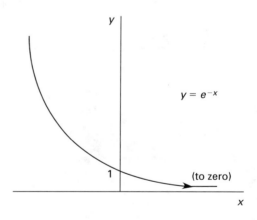

Natural Logarithmic Function

The *natural logarithmic function*, $f(x) = \ln x$, is defined as the *inverse* of the natural exponential function. That is, if $x = e^y$, then $y = \ln x$.

Since $x = e^y > 0$ for any real number y, it follows that the natural logarithmic function is defined *only for real numbers* $x > 0$. That is,

$$y = f(x) = \ln x \quad [x > 0]$$

Properties of $f(x) = \ln x$

Let x, y, and r be real numbers.

$$\ln e^x = x$$

$$e^{\ln x} = x \quad [x > 0]$$

$$\ln 1 = 0 \quad [\ln e = 1]$$

$$\ln (xy) = \ln x + \ln y \quad [x, y > 0]$$

$$\ln \left(\frac{x}{y}\right) = \ln x - \ln y \quad [x, y > 0]$$

$$\ln x^r = r \ln x \quad [x > 0]$$

$$-\infty < \ln x < \infty$$

Limiting Properties of $f(x) = \ln x$

$$\lim_{x \to 0^+} \ln x = -\infty$$

$$\lim_{x \to \infty} \ln x = \infty$$

Curve

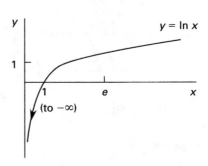

Index

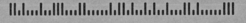